RAPID THERMAL PROCESSING FOR FUTURE SEMICONDUCTOR DEVICES

RAPID THERMAL PROCESSING FOR FUTURE SEMICONDUCTOR DEVICES

Proceedings of the 2001 International Conference on
Rapid Thermal Processing for Future Semiconductor Devices
(RTP 2001)
held at Ise-Shima, Mie, Japan,
November 14-16, 2001.

Edited by

Hisashi Fukuda

Muroran Institute of Technology
Department of Electrical and Electronic Engineering
27-1 Mizumoto-cho
Muroran, Hokkaido
050-8585 Japan

ELSEVIER

2003

Amsterdam – Boston – Heidelberg - London – New York – Oxford
Paris - San Diego – San Francisco – Singapore - Sydney – Tokyo

ELSEVIER B.V.
Sara Burgerhartstraat 25
P.O. 211, 1000 AE Amsterdam
The Netherlands

ELSEVIER Inc.
525 B Street, Suite 1900
San Diego, CA 92101-4495
USA

ELSEVIER Ltd
The Boulevard, Langford Lane
Kidlington, Oxford OX5 1GB
UK

ELSEVIER Ltd
84 Theobalds Road
London WC1X 8RR
UK

First edition 2003
Second impression 2003

Library of Congress Cataloging in Publication Data
A catalog record is available from the Library of Congress.

British Library Cataloguing in Publication Data
A catalogue record is available from the British Library.

ISBN: 0 444 51339 6

∞ The paper used in this publication meets the requirements of ANSI/NISO Z39.48-1992 (Permanence of Paper). Printed in Hungary.

Preface

This volume is a collection of papers which were presented at the 2001 International Conference on Rapid Thermal Processing (RTP 2001) held at Ise Shima, Mie, on November 14-16, 2001. This symposium is 2nd conference followed the previous successful 1st International RTP conference held at Hokkaido in 1997. The RTP 2001 covered the latest developments in RTP and other short-time processing continuously aiming to point out the future direction in the Silicon ULSI devices and II-VI, III-V compound semiconductor devices.

The first day of the symposium covered the future trend of shallow junction technology, advance MOS integration and novel metallization technology. Two invited papers are presented in the session related to future technology trend of RTP toward 21st century and single wafer process in DRAM manufacturing.

The second day of the symposium is the presentation of novel TFT fabrication, process monitoring and thin dielectric film formation technologies including two invited papers related to TFT fabrication and ferroelectric thin film formation. Poster session is covering the novel MOCVD processing, junction technology, novel oxidation, optoelectronics application etc.

The symposium ended with a session of novel RTP equipment and epitaxial thin film technologies, which included the invited paper related to wafer floating processing, and contributed papers for SiGe epitaxial process for HBT and quantum devices.

The symposium was sponsored by the 131st University-Industry Cooperative Research Committee on Thin Films and the 154th University-Industry Cooperative Research Committee on Semiconductor Interfaces and Their Applications organized in Japan Society for the Promotion of Science (JSPS). The symposium was successful from many aspects. Friendships were successfully formed through fruitful discussions and the social activities including a welcome-reception and banquet. The success of the symposium was greatly dependent on the efforts of organizing, steering and program committees, and also financial support of 10 organizations.

Y. Yasuda

Yukio Yasuda
Chairperson of RTP 2001

RTP2001 Organization

Chairperson: Y. Yasuda (Nagoya Univ.)
Vice chairperson: J. Murota (Tohoku Univ.)

Steering Committee:
S. Zaima (Nagoya Univ.)
A. Agarwal (Axcelis Technol.)
J. Hata (Tokyo Electron Ltd.)
S. Hirokawa (Canon Sales Co.,Inc.)
I. Kawai (Applied Materials Japan Inc.)
Y. Mimura (Ushio Inc.)
O. Takai (Nagoya Univ.)
L. Yasue (Dainippon Screen MFG. Co.,Ltd.)
K. Yoneda (Matsushita Electronics Corp.)

Program Committee:
H. Fukuda (Muroran Inst. of Technol.)
A. Agarwal (Axcelis Technol.)
K. Ando (NEC Corp.)
Y. Kunii (Hitachi Kokusai Electric Inc.)
S. Miyazaki (Hiroshima Univ.)
Y. Nakata (Altedec Co.,Ltd.)
N. Shah (Axcelis Technol.)
K. Suguro (Toshiba Corp.)
K. Takasaki (Fujitsu Lab.,Ltd.)
Y. Tanabe (Hitachi,Ltd.)
J. Tsuchimoto (Mitsubishi Electric Corp.)

Advisory Committee:
Y. Shiraki (Univ. of Tokyo)
A. Hiraki (Kouchi Inst. of Technol.)
T. Hara (Hosei Univ.)

Supported Organizations:
Applied Materials Japan Inc.
ASM Japan K.K.
Canon Sales Co.,Ltd.
Hitachi Kokusai Electric Inc.
Koyo Thermo Systems Co.,Ltd.
Mattson Technology Japan
NEC Corp.
Sumitomo Heavy Industries,Ltd.
Universal Systems Co.,Ltd.
Ushio Inc.

RTP2001 Secretariat:
K. Fukozu (SIPEC Corp.)
N. Yamamoto (SIPEC Corp.)

CONTENTS

Rapid Thermal Processing for Future Semiconductor Devices
H. Fukuda (Editor)

Role of Rapid Thermal Processing in the Development of Disruptive and Non-disruptive Technologies for Semiconductor Manufacturing in the 21st Century

Rajendra Singh †, Mohammed Fakhruddin and Kelvin F. Poole

Center for Silicon Nanoelectronics, and Holcombe Department of Electrical and Computer Engineering,
Clemson University, Clemson, SC 29634-0915, U. S. A.
† Phone: 1-864-656-0919 Fax: 1-864-656-5910 E-mail: srajend@clemson.edu

1.Introduction

Rapid thermal processing, popularly known as RTP has been around for more than two decades. In the context of silicon integrated circuit manufacturing, the equipment manufacturers and R&D groups from chip manufacturers as well as from universities have demonstrated a number of advantages over conventional furnace processing. As of today, RTP has not penetrated deep in Silicon IC manufacturing as it was forecasted and predicted by industry experts. Other than silicon ICs, the market size is very small. From the business point of view, the current status of RTP market is not very surprising since we operate in a world where only one out of 10 predicted markets ever come to fruition [1]. The introduction of 300 mm wafers in manufacturing (potential of 450 mm diameter wafer in future manufacturing), 70 nm gate length transistors in current 130 nm silicon integrated circuit manufacturing, the announcement of industrial researchers [2-3] for fabricating 15-20 nm gate length transistors, non-silicon IC technologies (display, storage, photovoltaics, optoelectronics etc.) and the issue of bringing disruptive technologies in production are some interesting issues that need to be considered for overall future of RTP. Thus, it is important to examine RTP technology from the current and future semiconductor manufacturing needs. In this paper, we have attempted to predict the future role of RTP in disruptive and non-disruptive semiconductor technologies.

2.Global Picture of Semiconductor Manufacturing

An examination of the growth of the semiconductor industry shows that the ability to reduce the feature size (leading to lower cost/per function) is largely responsible for the sustained growth in the last four decades. The market size in 1959 was less than $1 B and the current market size is about $200 B. In spite of several negative growth cycles, over the last 40 years the semiconductor industry has maintained 17 % compound annual growth rate [4]. The overall semiconductor market consists of silicon ICs for low-power applications as well as for power-electronics, micro electro mechanical systems (MEMS), compound semiconductor ICs, optical devices and optoelectronics, flat panel display, solar cells as well as disruptive technologies. In principle RTP has potential to meet thermal processing and chemical vapor deposition (CVD) needs of all the above-mentioned technologies. However, our discussion in this paper is focused mainly on silicon ICs for low power applications and disruptive technologies. In a recent paper we have discussed the role of RTP for photovoltaic applications [5].

In recent years researchers are raising an open question about how far the feature size reduction trends of silicon based CMOS will continue [6-7]. Alternate device structures (e.g. single electron transistors (SETs)), alternate materials (e.g. carbon nanotubes, magnetic materials, biological materials), molecular electronics, optical switches, quantum computers etc. have been proposed as an answer to the question, "what after silicon CMOS? ". Careful examination of all the published results show that none of the proposed solutions can lead to manufacturing. The best results published to date involve the use of self-assembled monolayers of conjugated polymers in a vertical CMOS structure [8]. Due to small channel length (`1-2 nm), the drain current is higher than the smallest silicon based CMOS reported to date. However, due the unstable nature of the organic semiconductor, the drain was deposited at 100K. Thus, from defects and reliability considerations [9], it is unlikely that the work reported in reference 8 can lead to manufacturing. The feature sizes of Si CMOS in 1980 and year 2000 were about 2 micron, and 0.2 micron respectively. It is expected that by the year 2020, the feature of Si CMOS will be about 20 nm.

3.Current Trends in Semiconductor Manufacturing

During the last 40 years, due to increased wafer size, decreased line widths and larger capacities in wafer out per month have driven a decrease in capital cost per normalized unit (e.g. cost per transistor of an IC chip). However, the cost of semiconductor facilities is increasing exponentially with time. In recent years, the most difficult challenge for IC manufacturer is to meet the demands of time-to-market of a particular product. The current mega fabs are dominated by batch processing. In future, batch processing will play a reduced role in mega fabs. The agile manufacturing based on mini fabs will be driven by shorter products life and dominated by single wafer processing. In the context of current manufacturing trends the key features of agile manufacturing are [10]: (i) driven by quick turn around time and low-cost, (ii) small size fab supported by a virtual fab, (iii) multi-task and muti-functional tools, (iii) minimization and smoothing of raw process time

Historically, the concept of agile manufacturing for microelectronics was introduced in 1989 [11]. Funded partially by US Department of Defense, Texas Instruments completed microelectronics manufacturing science and technology (MMST) project in 1993. Using home built single wafer processing tools, TI demonstrated a cycle time of 3 days. The details of earlier work of TI and the limits of single wafer processing are discussed in a recent paper by Doering and Nishi [12].

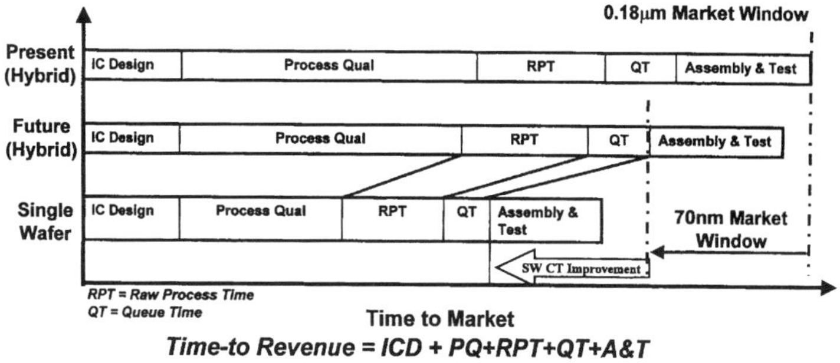

Fig.1 Single Wafer Processing shortens time-to-revenue and enables users to hit the accelerate cash inflow [14].

Currently the equipment manufacturers are divided on the issue of single wafer processing vs batch processing. Current vendors of thermal batch processing tools advocates that with innovation, the batch processing can meet the future needs of microelectronics manufacturing. Some of the innovative steps are: (i) small load size and high speed transfer, (ii) ultra fast heater and thermo plug combination valve, (iii) fast ramp down, soft back fill injector and small reactor volume, and (iv) rapid cooling units [13]. Although, the details of the approximations used in fab simulations are not known, it is claimed that after production wafer in progress reaches steady state levels, the batch processing actually provides shorter turn around time than single wafer processing [13]. As shown in Fig. 1 [14] the single wafer processing (SWP) vendors advocates that SWP shortens the time- to -revenue and enables users to hit the accelerate cash inflow.

In addition to the early work of Texas Instruments, currently there are many chip manufacturers who are advocating SWP over batch processing. As an example, Koike [15] has recently demonstrated the cycle time reduction features of SWP. These results are summarized in Table I. Our own general observation about cost vs raw processing time is that currently single wafer processing cost is generally 2-4 times higher than the batch processing. On the other hand, SWP raw processing time is about 2-4 times less than batch processing. For manufacturing beyond 70 nm node, we expect the dominance of SWP. In the ultimate limit of IC manufacturing, the SWP will take over batch processing.

4.Lessons from the History of RTP

It is important to look at the history of RTP. The first paper dealing with the use of low-thermal mass graphite heater for annealing GaAs ion-implanted wafers was reported in 1977 [16]. In 1980, Nishiyama and co-workers [17] used halogen lamps as a continuous source of radiation to anneal boron-implanted Si for a very short time (6 sec.). The use of incoherent lamps as the source of energy led to the field of rapid thermal processing. For the early history of RTP, one is refereed to Ref. [18]. In the 80s, the availability of R&D commercial tools by AG Associates was one prime reason that a number of researchers contributed in this field.

Table I Cycle Time (arbitrary unit) Simulation of 300 nm Fab [15]

	25 wfs/lot	13 wfs/lot	3 wfs/lot
Batch Process	43		
Single Wafer Process	25	17	10

Most of the researchers attributed low-thermal mass leading to high heating cooling rates as the only difference between RTP and conventional furnace processing (CFP). As early as 1985 [19], one of us pointed out the fundamental difference in the energy spectrum of resistive heater based CFP systems and incoherent lamp based RTP systems. We pointed out the reduction of cycle time and process activation energy as the two distinct advantages of RTP over CFP [20]. Temperature measurement, and temperature uniformity were identified as the major issues in the use of RTP in manufacturing. Although, termed as Cinderella Technology [21], RTP did not find any manufacturing application in 80s. Titanium silicidation appeared as the first use of RTP in manufacturing [22]. However, the introduction of better control systems in CFP, use of the mini-furnaces, and partial or no use of photo effects in RTP are the main reasons that RTP could not replace CFP as it was expected in earlier studies.

5.Key Issues for the Development of Future RTP Systems

The introduction of new interconnect materials (copper as well as low dielectric constant materials) has already allowed the continuous feature size reduction of silicon ICs. At the gate stack level, high-dielectric constant materials and new gate materials are absolute necessities for sub-50 nm node technologies. As shown in Fig. 2 [modified from Ref. 23], for cost-effective manufacturing (minimization of process steps) reduction of processing temperature is a very important process integration issue.

For the manufacturing of ultimate minimum feature size silicon ICs it is widely believed that alternate transistor structure (e.g. double gate CMOS) will be required. A comparison of single and double gate CMOS devices is given in Fig. 3 [24]. As shown in Fig. 4 [24], the use of current RTP tools does not provide minimum variance for double gate CMOS devices. Thus future RTP systems should provide reduced variance of the critical device design parameters.

In a recent publication [9], we have addressed the most important issues for semiconductor manufacturing in the 21st century For obtaining improved performance, reliability and yield of any semiconductor device, use of homogenous microstructure is a necessity. This can be achieved by optimizing the thermal cycle for obtaining minimum and uniform thermal stress. Fig. 5 depicts the essential features of our approach in obtaining homogenous microstructure materials. A complete details of the process outlined in Fig. 5 is given in Ref. [25] and other related publications [26-28].

6.New RTP Markets

From an equipment manufacturer's point of view (different chemistries provide a number of applications), all the current and future applications of RTP can be divided in two categories: (i) single wafer substrates, and (ii) multi-wafer or larger size substrates. The first category is the driver of current applications of RTP. The introduction of new materials in manufacturing is possible when RTP can provide significant advantages over other processing technologies. As an example, the introduction of high dielectric constant as the gate dielectric material can provide ultimate Si CMOS devices for manufacturing. As pointed out in Ref. [2], the sub-threshold slope of current devices is 100 mV/decade. If the RTP based CVD systems can provide a value close to 60-70 mV/decade, a whole new CVD based market will open. This will require a change in the current practice. In addition to Si ICs, basically the same equipment (with change of chemistry and some other details) can open new markets for compound semiconductor processing.

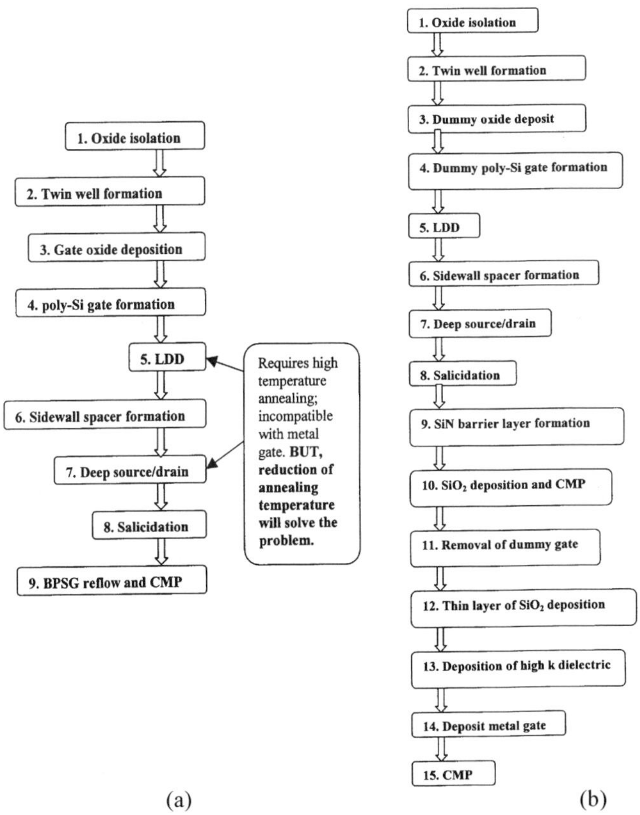

Fig.2 CMOS process flow: (a) Standard process flow (b) Proposed [23]-process flow, using dummy gate.

The second category involves areas such as solar cells, and display technologies. The tolerance for temperature variations is greater than what is required for applications discussed earlier. However, these applications require order of magnitudes higher throughputs. By significant reduction of activation energy and using lower-cost control systems (compared to single wafer based RTP) these applications are possible.

7.Conclusions

In this paper we have analyzed the current and future prospects of RTP for semiconductor manufacturing. Current trends are in the direction of single wafer processing based manufacturing and RTP will play a very important role. We have pointed out the essential features that need to be introduced in the design of future RTP systems. Specifically, reduction of processing temperature, reduction of variance of critical device parameters and obtaining homogenous microstructures are the key issues for designing future RTP systems.

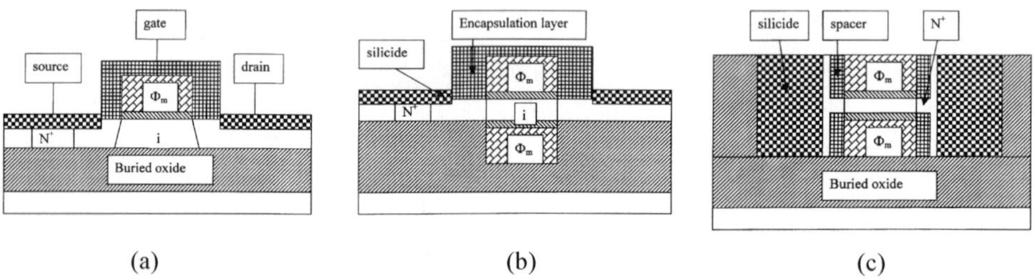

Fig.3 Schematics of Single Gate (SG) and Double Gate (DG) devices: (a) an ultra thin body SOI single-gate device; (b) the same device with two gates; (c) an ideal double-gate device [24].

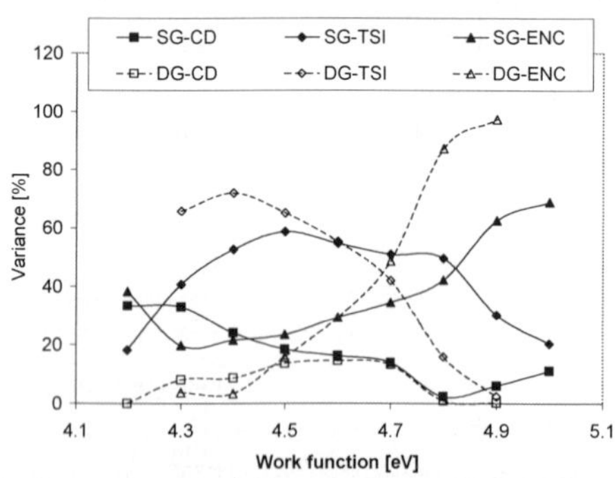

Fig.4 Study of Variance of I_{off} for 40 nm single gate (SG) and double gate (DG) technologies [24]. The three dominant sources of variations are silicon thickness (TSI), critical dimension (CD), and encapsulating spacer thickness (ENC) control.

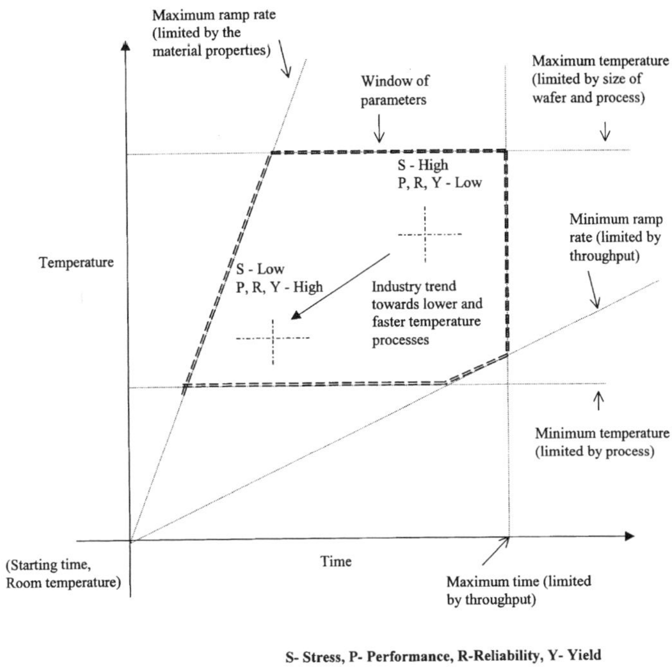

Fig.5 Optimization of thermal cycle in RTP

Acknowledgements

One of us (RS) would like to thank former students, research associates and a number of colleagues in industry and academia. Special thanks to Drs. Chu, Thakur, Shrangpani, Chen and Parihar whose research contributions provided the framework of this work. The authors would like to thanks Anthony Dip and Sanjeev Kaushal of Tokyo Electron America, Zoran Kirvokapic of AMD, Kamal Rajkanan of KLA-Tencor, Randhir Thakur of Applied Materials, and Masahiro Yoshimura of Tokyo Institute of Technology for providing their pre-published research results. NSF, DARPA, ATP, ARO and AG Associates supported this work.

References

[1] M. Moritz, *Forbes ASAP*, October 2, 2000, 41.

[2] B. Yu, H. Wang, A. Joshi, Q. Xiang, J. Jeon, and M. Lin, Late News Paper, *Int. Electron Devices Meeting,* December 2001.

[3] Intel Researchers Build World's Fastest Silicon Transistors, June 11,2001 http://www.intel.com/pressroom/archive/releases/20010611tech.htm

[4] M. S. Malone, *Forbes ASAP*, September 10, 2001, 44

[5] R. Singh, A. Rohatgi, K. Rajan and K. F. Poole, *Electrochemical Society Proc.* Vol. # PV 99-11, (1999), 33.

[6] P. A. Packan, *Science*, 285, (1999) 2079.

[7] M. Schultz, *Nature*, 99, (1999) 729.

[8] J. H. Schon, H. Meng and Z. Bao, *Science* 294 (2001), 2138.

[9] R. Singh, V. Parihar, K. F. Poole and K. Rajkanan, *Semiconductor Fabtech*, 9thEdition (1999) 223

[10] Y. Mikata, K. Mitsutake, T. Arikado, and K. Okumura, *Conf. Proc. IEEE Int. Symp. on Semiconductor manufacturing,* (2001), 7.

[11] K. Saraswat, *SPIE Rapid Isothermal Processing Conf. Proc.* vol. 1189, edited by R. Singh (1989) 1.

[12] R. Doering and Y. Nishi, *Proc. IEEE*, 89 (2001) 375.

[13] Anthony Dip, Tokyo Electron America, personal communication, October 2001.

[14] P. Meissener, Workbook of Workshop on Manufacturing Issues Beyond the 70 nm Node, ISSM, October 2001

[15] A. Koike, *VLSI Techno. Symp.*, 2001, 1.

[16] R. K. Surridge, et al., *Proc. Int. Symp. GaAs and Related Compounds*, **33a**, (1977) 161.

[17] K. Nishiyama, M. Ari, and N. Watanabe, *Jpn. J. App. Phys.*, **19**, (1980) L563.

[18] R. Singh, *J. App. Phys.* **63** (1988), R59.

[19] R. Singh, *Electronics* **58** (1985), 19.

[20] R. Singh, *Semiconductor Int.* **9(1)** (1986) 28.

[21] P. H. Singer, *Semiconductor Int.* **12(4)** (1989) 28.

[22] P. H. Singer, *Semiconductor Int.* **16(6)** (1993) 64.

[23] K. Matsuo, T. Satio, A. Yagishita, T. Iinuma, A. Murakoshi, K. Nakajima, S. Omoto, and K. Suguro, *Symp. on VLSI Tech*, (2000) 70.

[24] Z. Krivokapic, and W. D. Heavlin, *IEEE Trans. on Semiconductor Manufacturing*, to be published (May 2002).

[25] R. Singh, V. Parihar, Y. Chen, K.F.Poole, S. Nimmigadda, and L. Vedula, *IEEE Trans. Semiconductor Manufacturing,* **12** (1999), 36.

[26] R. Singh, Chap. 9, Hand Book of Compound Semiconductors, Noyce Pub. (1995) 442.

[27] R. Singh, and V. Parihar, Vol. 2, Chap. 1, *Hand Book of Low and High Dielectric Constant Materials*, Academic Press (1999) 1.

[28] R. Singh, M. Fakhruddin and K. F. Poole, *Applied Surface Science*, **168** (2000) 198.

Rapid Thermal Processing for Future Semiconductor Devices
H. Fukuda (Editor)

Analytical Model for Spike Annealed Diffusion Profiles of Low-Energy and High-Dose Ion Implanted Impurities

Kunihiro Suzuki and Hiroko Tashiro
Fujitsu Laboratories Ltd.
10-1 Morinosato-Wakamiya Atsugi 243-0197, Japan
Tel: +82-463-250-8236, E-mail: suzuki@flab.fujitsu.co.jp

1.Introduction

Scaling theory requires shallow junctions [1-4]. Spike annealing, where the isothermal time period is zero or quite short, enables us to obtain these shallow junctions. Combined with low-energy ion implantation, this technology is hence frequently used in the high-speed MOSFET processes [5]. Temperature is always time dependent in these processes, and we need to acquire accurate parameters over a wide temperature range including the transient effects of the processes to analyze these processes. This kind of rigorous treatment has been performed in commercial process simulators [6-8]. However, many ambiguous models and their related parameters exist. Therefore, it is difficult to obtain a clear idea of how to optimize the spike annealing conditions.

Many researchers have studied high concentration diffusion profiles using analytical approaches [9-16]. However, the diffusion coefficient depends on the doping concentration [17], and the diffusion equation thus becomes nonlinear and difficult to solve. Nakajima and Fair proposed models for the profiles [9-12]. These models are functions to express the profiles and are not based on the diffusion equation. Anderson solved the nonlinear equation in the case of a constant total dose condition [13, 14]. However, the common boundary condition for the high concentration diffusion is the constant surface concentration instead of the constant total dose in high dose conditions. We derived a diffusion model with the constant surface concentration based on the nonlinear diffusion equation [15, 16].

Here, we extend our model for high-concentration-diffusion profiles to accommodate spike annealed diffusion profiles and the corresponding junction depth, x_j, and show the optimum condition to obtain shallow the junctions with abrupt profiles.

2.Experimental

We used Si(100) substrates, and ion implanted B at 0.5, 1, 3, 5 keV or As 1, 3, 5 keV and a dose of 1×10^{15} cm^{-2}. The substrates were then subjected to rapid thermal annealing with the ramp-up rate of 60°C/sec and isothermal process time was 1 sec. The impurity profiles were evaluated using secondary ion mass spectrometry (SIMS). The junction depth x_j was defined by the depth where impurity concentration is 10^{18} cm^{-3}.

3.Results

Figure 1 shows B SIMS diffusion profiles at the energies of 0.5, 1, and 3 keV for the maximum temperatures of 975, 1000, and 1025°C. At the energy of 0.5 keV, x_j decreased with decreases of the maximum temperature as was expected. However, x_j became insensitive to the maximum temperature with increases of the acceleration energy, and finally, it became invariant at 3 keV. This is because transient enhanced diffusion (TED) dominates x_j, and TED starts and finishes during the ramp-up time period before reaching 975°C [17]. The thermal equilibrium diffusion dominates x_j at low energy since TED is suppressed in the low-energy region [18-20]. x_j then becomes dependent on the annealing condition. On the other hand, x_j is less dependent on acceleration energy in the low energy range of less than 1 keV. Therefore, it is good to reduce energy to be free of TED and control x_j by temperature.

Figure 2 shows As SIMS diffusion profiles at the energies of 1, 3, and 5 keV for the maximum temperatures of 975, 1000, 1025°C. x_j always decreased with decreases of the maximum temperature for all energy ranges as was expected. This is because As is insensitive to TED.

Figure 3(a) summarizes the dependence of x_j associated with B diffusion profiles for the maximum annealing temperatures. Dashed lines correspond to the x_j of as-implanted profiles. In the low-energy region of less than 1 keV, x_j increases with increases of temperature, and the difference of x_j between 0.5 and 1 keV becomes at the increased maximum temperature. This is because the diffusion length is much deeper than the x_j of the as-implanted profiles. x_j is independent of temperature in the higher energy range of 3 to 5 keV. This is because TED dominates x_j, as mentioned before.

Figure 3(b) summarizes the dependence of x_j of the As diffusion profiles on the maximum annealing temperature. Dashed lines correspond to x_j of as-implanted profiles. x_j increased with increases of temperature over the whole acceleration energy range as was expected.

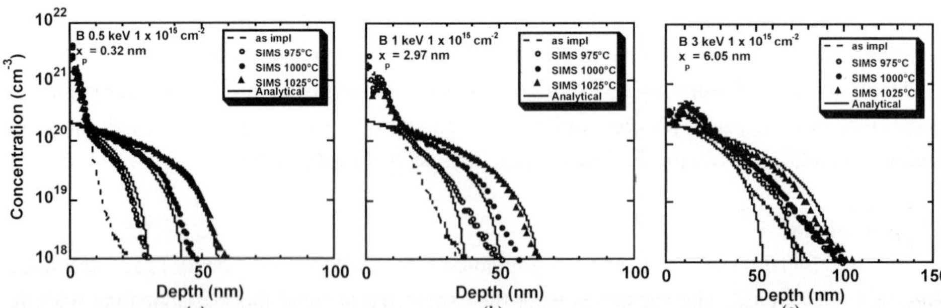

Fig.1 Boron diffusion profiles. Analytical models are also shown. (a) 0.5 keV, (b) 1 keV, (c) 3 keV

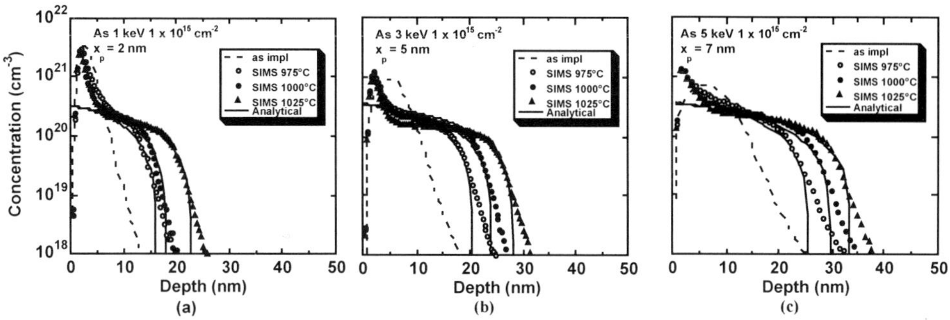

Fig.2 Arsenic diffusion profiles. Analytical models are also shown. (a) 1 keV, (b) 3 keV, (c) 5 keV.

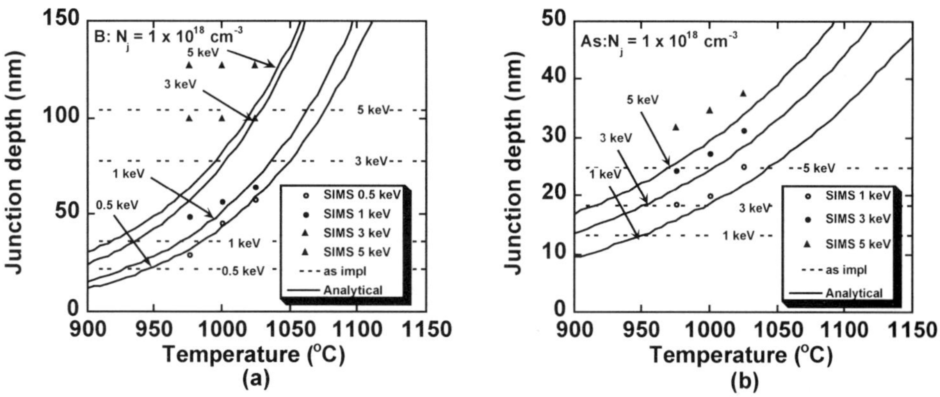

Fig.3 Summary of junction depths of impurity diffusion profiles. (a) B, (b) As.

4.Discussion

We developed a model for high-concentration-impurity diffusion profiles [15] given by

$$N = N_s\left(1 - \sqrt{2\gamma}Y - Y^2\right)^{1/\gamma} \tag{1}$$

where N_s is the maximum diffusion concentration, Y is the normalized depth, and γ is the parameter which expresses the significance of the impurity diffusion coefficient on concentration and is 0.5 for B and 1 for As [16]. Y is expressed by

$$Y = \frac{x}{\sqrt{\dfrac{4(\gamma+2)\eta_{\gamma a}}{\gamma}\left(\dfrac{N_s}{n_i}\right)^{\gamma} D_i t}} \tag{2}$$

where $\eta_{\gamma a}$ is a constant number and depends on γ, however it is around 1.7 [15]. Equation 1 is valid for the isothermal process. In the spike annealing processes, it is unclear what are D_i and t in our model. It may be possible to include the time dependent N_s, n_i, and D_i in the model. However, we need to have accurate values for N_s, n_i, and D_i over a wide temperature range and accurate time-temperature curves, which are very difficult to obtain. Therefore, we adopt an empirical approach with the aid of experimental data.

The denominator of Y in Eq. 1 is related to the diffusion length, and we express the denominator as x_0, therefore, Y is then expressed by

$$Y = \frac{x - x_p}{x_0\left(T_m, E\right)} \tag{3}$$

x_p is the depth at the maximum concentration, and we use the projected range R_p of ion implantation for x_p. x_0 empirically expresses the diffusion length that should be related to the ramp-up, ramp-down, and T_m. Inspecting Eqs. 2 and 3, x_0 is a function of N_s, n_i, and D_i, and they are all expressed by their corresponding activation energies. Therefore, we expect that x_0 is expressed by an activation energy, which is t be verified later.

x_j can be estimated as N = 0 in Eq. 1. The corresponding Y, Y_j, is given by

$$Y_j = \frac{\sqrt{2}}{\sqrt{\gamma} + \sqrt{\gamma + 2}} \tag{4}$$

and hence x_j is given by

$$x_j = x_p(E) + \frac{\sqrt{2}}{\sqrt{\gamma} + \sqrt{\gamma + 2}} x_0\left(T, E\right) \tag{5}$$

The definition of x_j is the difference between the analytical model and the experimental data. However, the profile based on the analytical model is abrupt near the junction, and hence the value is insensitive to the concentration. We extracted x_0 so that Eq. 1 fits the SIMS data in the high-concentration region shown in Fig. 4, and they are given by

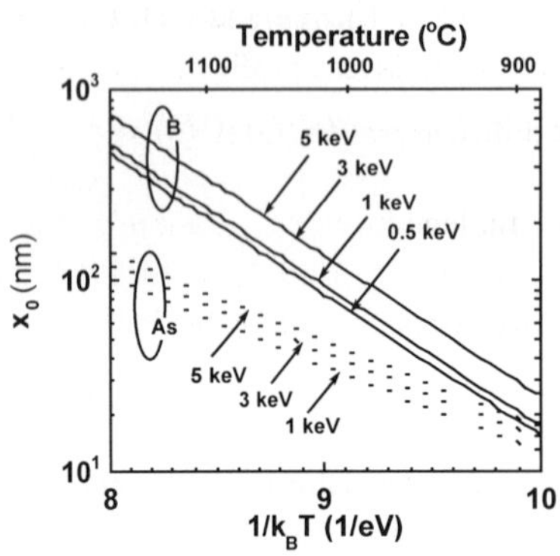

Fig.4 Dependence of extracted diffusion length on temperature

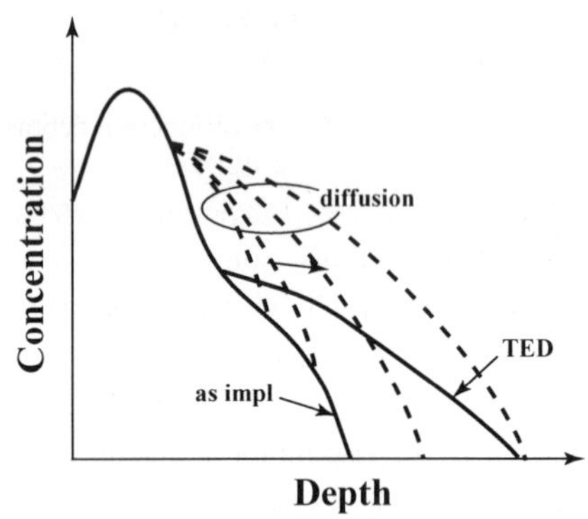

Fig.5 Schematic diffusion profiles. TED starts and finishes before normal diffusion occurs. The normal diffusion proceeds in the high concentration region, and the junction depth is invariant before the normal diffusion reaches the junction depth associated with the as-implanted profile or TED.

$$x_0(B) = \begin{cases} (3.3 + 0.9E) \times 10^8 \exp\left[-\dfrac{1.7(eV)}{k_B T}\right] nm & for\ E \leq 3\ keV \\ 6 \times 10^8 \exp\left[-\dfrac{1.7(eV)}{k_B T}\right] nm & for\ E > 3\ keV \end{cases} \tag{6}$$

$$x_0(As) = (4.2 + 0.4E) \times 10^5 \exp\left[-\dfrac{1.05(eV)}{k_B T}\right] nm \tag{7}$$

x_0 depends on temperature and energy. x_0 for As is smaller than that for B, which is plausible since the diffusion length of As is smaller than that of B. x_0 weakly depends on the energy value. We do not know what is the physical meaning of the energy dependence.

The analytical model agrees well with the experimental data for 0.5 keV for the B diffusion profiles. Our model reproduces the experimental data in the high-concentration region for the higher energy regions. However, it deviates from the experimental data in the low concentration region. The profiles suffer from TED and our model does not include TED. The analytical model agrees well with the experimental data in the high-concentration region of the As diffusion profiles. However, the experimental data deviate from the analytical model in the low-concentration region. The experimental data may suffer resolution limit of SIMS measurement. However, we do not know its significance, or if the tail profile in the low-concentration region is real or not.

Our model focuses on the profiles associated with normal diffusion and assumes that the initial junction depth is x_p, which is the peak position of as-implanted profiles, and it does not include TED. Therefore, the information associated with as-implanted profiles and TED should be obtained experimentally.

We denote the junction depth associated with the as-implanted profile x_j(as-impl), that with TED x_j(TED) and that with the thermal equilibrium diffusion x_j(diff). TED starts and finishes within quite a short time, and the thermal equilibrium diffusion occurs dominantly in the high-concentration region and x_j(as-impl) and x_j(TED) are invariant during the thermal equilibrium diffusion, as shown by the schematic in Fig. 5. Therefore, each junction depth can be treated independently and the final junction depth is the maximum value among the junction depths, expressed by

$$x_j = Max\left[x_j(as\ impl),\ x_j(TED),\ x_j(diff)\right] \tag{8}$$

Note that the abruptness of the diffusion profile is better than that of as-implanted or TED profiles. Therefore, we can regard the optimum profile as the normal diffusion profile that dominates x_j. Therefore, the diffusion condition should be selected so that thermal equilibrium diffusion dominates x_j, that is

$$Max\left[x_j(as\ impl),\ x_j(TED),\ x_j(diff)\right] = x_j(diff) \tag{9}$$

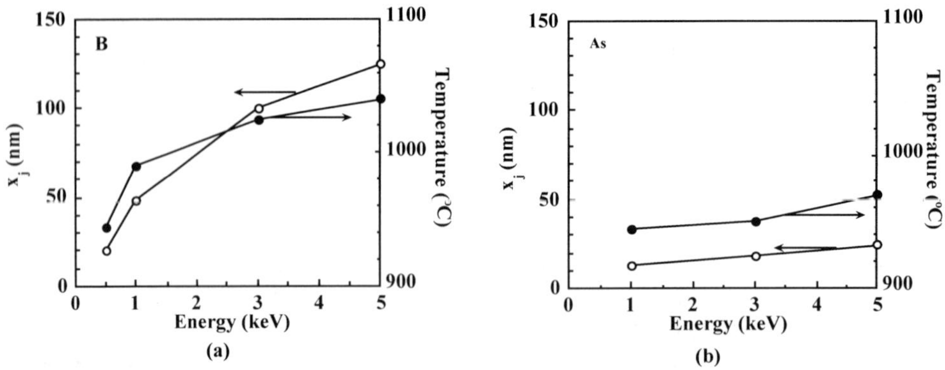

Fig.6 Dependence of optimum junction depth on acceleration energy. The corresponding maximum temperature is also shown.(a) B, (b) As.

Figure 3(a) shows the analytical model for the dependence of x_j on maximum temperature for B diffusion profiles. We can draw lines that correspond to x_j associated with normal diffusion. We can estimate the optimum diffusion condition graphically as a cross point of the line and junction depth associated with the as-implanted profiles or TED profiles. For 0.5 keV, the optimum point was the cross point of the junction depth associated with the as-implanted profile and that with the normal diffusion profile. For 1 keV, we observed TED at 975°C, and the optimum point was the cross point of TED and the normal diffusion. For 3, and 5 keV, the optimum point clearly was the normal diffusion condition where the diffusion reached x_j associated with TED.

Figure 3(b) shows the analytical model for the dependence of x_j on the maximum temperature for the As diffusion profiles. The optimum diffusion condition is the cross point of the line of x_j associated with the normal diffusion and that with as-implanted profiles, as TED does not affect the diffusion profiles significantly for As diffusion.

Figure 6 summarizes the optimum junction depth. As the energy decreases, the junction depth decreases accordingly with lower maximum temperatures. For 0.5 keV B, a 20-nm junction depth can be expected at 945°C spike annealing, and 13 nm junction depth can be expected for As 1 keV at 945°C.

5.Summary

We showed systematic spike annealed diffusion data and showed that low-energy B ion implantation is effective to be free from transient enhanced diffusion. We extended our high-concentration diffusion model to spike annealing profiles by introducing an effective diffusion length. Our model well expresses the diffusion data in the high-concentration region, but deviates from the data in the low-concentration region, where transient enhanced diffusion or the as-implanted profile dominates the junction depth. Optimum points are evaluated as the cross point of the junction associated with the normal diffusion junction and junctions associated with the as-implanted profiles or the transient enhanced diffusion. We can realize 20-nm junction depth with 0.5 keV and 945°C spike annealing for B, and 13 nm junction depth with 1 keV and 945°C spike annealing for As.Since our model is empirical, we should obtain similar data related to annealing equipment used, and corresponding the ramp-up period and ramp-down process.

References

[1] R. H. Dennard, F. H. Gaensslen, H. N. Yu, V. L. Rideout, E. Bassous, and A. L. Blanc, "Design of ion-implanted MOSFETs with very small physical dimensions, " IEEE J. Solid-State Circuits, vol. SC-9, pp. 256-268, 1974.

[2] J. R. Brews, W. Fichtner, E. H. Nicollian, and S. M. Sze, "Generalized guide for MOSFET miniaturization, " IEEE Electron Dev. Lett., vol. EDL-1, pp. 2-4, 1980.

[3] P. Chatterjee, W. R. Hunter, T. C. Holloway, and Y. T. Lin, "The impact of scaling laws on the choice of n-channel or p-channel for MOS VLSI, " IEEE Electron Dev. Lett., vol. EDL-1, pp. 220-223, 1980.

[4] G. Baccarani, M. R. Wordeman, and R. H. Dennard, "Generalized scaling theory and its application to a 1/4 micrometer MOSFET design," IEEE Trans. Electron Dev., vol. ED-31, pp. 452-462, 1984.

[5] A. Agarwal, H. J. Gossmann, and A. T. Fiory, Tech. Digest of Solid-State Devices and Materials, p. 508, 1999.

[6] TSUPREM IV USERS' manual, AVANT!.

[7] SSUPREM IV USERS' manual, SILVACO.

[8] DIOS USERS' manual, ISE.

[9] Y. Nakajima, S. Ohkawa, and Y. Fukukawa, "Simplified expression for the distribution of diffused impurity," Japan J. Appl. Phys., vol. 10, pp. 162-163, 1971.

[10] R. B. Fair and J. C. C. Tsai, "The diffusion of ion-implanted arsenic in silicon," J. Electrochem. Society, vol. 122, pp. 1689-1696, 1975.

[11] R. B. Fair and J. C. C. Tsai, "Profile parameters of implanted-diffused arsenic layers in silicon," J. Electrochem. Society, vol. 123, pp. 583-586, 1976.

[12] R. B. Fair, "Boron diffusion in silicon-concentration and orientation dependence, background effects, and profile estimation," J. Electrochem. Society, vol. 122, pp. 800-805, 1975.

[13] D. Anderson and M. Liasak "Approximate solutions of some nonlinear diffusion equations," Physical Review A, vol. 22, pp. 2761-2768, 1980.

[14] D. Anderson and K. O. Jeppson, "Nonlinear two-step diffusion in semiconductors," J. Electrochem. Society, vol. 131, pp. 2675-2679, 1984.

[15] K. Suzuki, "High concentration impurity diffusion model," Solid-State Electronics, vol. 44, pp. 457-463, 2000.

[16] K. Suzuki, "High concentration diffusion profiles of low energy ion-implanted B, As, and BF2 in bulk Silicon," Solid-State Electronics, vol. 45, pp.1747-1751, 2001.

[17] S. K. Ghandhi, VLSI Fabrication Principles, John Wiley&Sons, Inc., New York, 1994.

[18] K. Suzuki, M. Aoki, Y. Kataoka, N. Sasaki, A. Hoefler, T. Feudel, N. Strecker, and W. Fichtner, "Analytical models for transient diffusion and activation of ion-implanted boron during rapid thermal annealing considering ramp-up period," International Electron Device Meeting, pp.799-802, 1996.

[19] K. Suzuki, T. Miyashita, Y. Tada, A. Hoefler, N. Strecker, and W.Fichtner, "Damage calibration concept and novel B cluster reaction model for B transient enhanced diffusion over thermal process range from 600°C (839 h) to 1100 (5s) with various Ion implantation doses and energies," International Electron Device Meeting, pp.501-504, 1997.

[20] A. Agarwal, H. J. Gossmann, D. J. Eaglesham, L. Pelaz, D. C. Jacobson, T. E. Haynes, and Y. E. Erokhin, Appl. Phys. Lett., vol. 71, p. 3141, 1997.

Rapid Thermal Processing for Future Semiconductor Devices
H. Fukuda (Editor)

Process and Technology Drivers for Single Wafer Processes in DRAM Manufacturing

R. A. Weimer[1], D. C. Powell[1], and P. M. Lenahan[2]

[1]Micron Technology, Inc., 8000 S. Federal Way, MS 306, Boise, ID 83707-0006, USA
Phone: 208-368-5605 Fax: 208-368-2548 E-mail: rweimer@micron.com
[2]The Pennsylvania State University, University Park, PA 16802, USA
Phone: 814-863-4630 E-mail: pmlesm@engr.psu.edu

1.Introduction

Cost of ownership (COO) continues to determine that batch wafer systems dominate front-end-of-the-line (FEOL) DRAM processes such as low pressure chemical vapor deposition (LPCVD) of films, oxidation cycles, and wafer anneal cycles. As of today, process advantage for a single wafer technology can only be gained by an enabling process that is not available in a batch system. Therefore, to displace the batch system, single wafer technologies must show some unique process capability or significant process advantage. As DRAM devices shrink below the 0.15μm technology node, however, the process control of the transistor devices becomes more critical for optimal DRAM performance. Thus, one potential process advantage may be gained by a reduced thermal budget for single wafer tools in process areas still dominated by batch processing. In this paper, we review the current status of single wafer FEOL processes in DRAM production as well as possible paths for current batch processes to move toward single wafer tools.

2.Current Rapid Thermal Processing (RTP) Anneal Technologies

As shown in Table I, each RTP anneal technology has potential advantages and disadvantages for processing wafers. In choosing an RTP tool set for DRAM manufacturing, cost of ownership and tool footprint are two very important parameters to consider. As reviewed in previous work, DRAM devices to date have not needed the advanced temperature control and temperature uniformity that logic devices need due to the high thermal processing associated with construction of the DRAM cell capacitor (for non-trench cells) [1]. As projected, however, the need for high-temperature uniformity is close at hand due to decreasing gate lengths in DRAM technology.

Table I. Brief review of the current commercially available RTP technologies for BPSG reflow and spike anneal.

Technology (heating/temperature control)	Advantages	Possible issues
Axi-symmetric lamps/ reflecting plate	Ease of rotation No interference between temperature control and wafer MIMO temperature control	Coating of reflecting plate More susceptible to pattern effect Power requirements Lamp/quartz interactions
Double-sided lamps/ripple pyrometer™	Known technologies Reduced complexity Reduced pattern effect	Coating of quartz ware/temperature control No MIMO
Furnace core/augmented pyrometer	Very simple Low power Little pattern effect	No MIMO Uniformity changes very difficult
Isothermal heater(s) front and/or back	Very simple Low power No pattern effect	Emissivity issues on ramp Uniformity cannot be changed

Most of the temperature control systems in RTP systems as of today can safely claim <4°C total temperature variation wafer-to-wafer as a result of wafer emissivity change. As shown in Fig. 1, this particular technology has a 4°C variation due to backside emissivity changes and an approximate 3°C offset with variation of front side films. However, even the most advanced temperature control systems can have issues resulting from chamber conditioning. As shown in Fig. 2, even with a relatively simple process such as oxide growth, the most advanced RTP technologies may suffer from mundane issues such as first wafer effect. Another area of continuing emphasis should be in the area of tool-to-tool matching. Although temperature control on individual tools is excellent, process variation tool-to-tool may be more than the individual chamber variation specifications, depending on the calibration-matching technique for the given RTP technology.

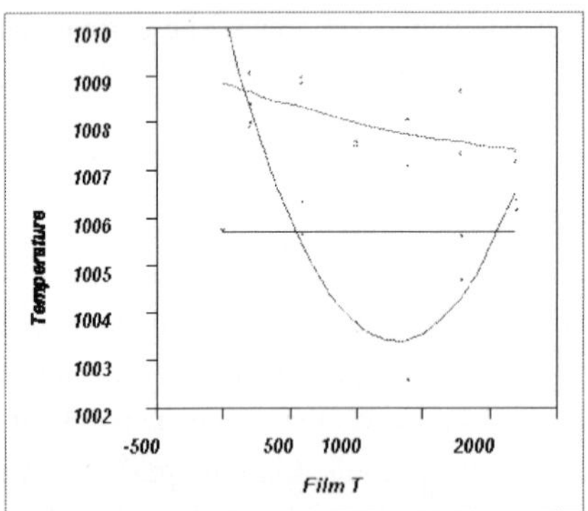

Fig.1 Emissivity testing of an RTP technology with bare silicon (blue),emissivities ranging from 0.36–0.6 (red), and front side emissivities (0.36–0.6) (green).

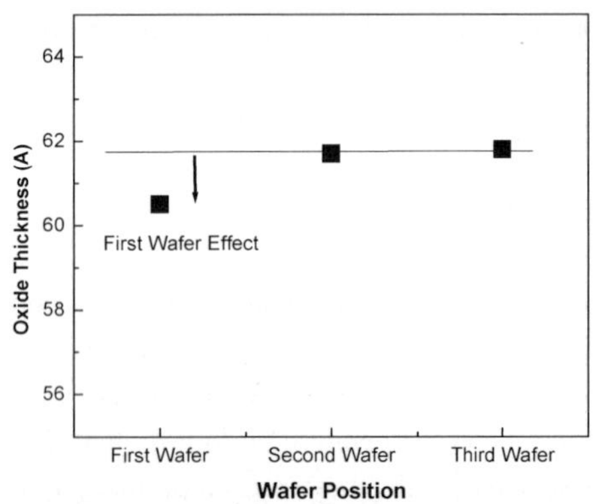

Fig.2 Oxide growth in RTP system showing first wafer effect even with advanced RTP temperature control.

3.BPSG Reflow

RTP technology has now been in production for doped-glass reflow for over ten years. The need for high-temperature reflow of doped oxide is due to the very tight wordline-to-wordline pitch in a DRAM device relative to that of a logic device. Inadequate reflow can result in shorts between the wordlines or digit contacts. Due to the high temperatures needed to reflow the doped glass, a batch system would densify the BPSG before the reflow could be achieved. As shown in previous publications, this RTP anneal is typically the first major heat step post-transistor implants and thus can account for up to 30% of the final transistor variation [1]. The temperature uniformity of this anneal has not been a critical issue in the past. As mentioned, however, as DRAM technologies continue to shrink, the temperature uniformity of the doped glass reflow will become more critical.

4.Gate Dielectrics

One of the potential future applications for single wafer processing for DRAMs would be gate-oxidation. It is known that oxides grown with steam (and usually some post-treatment or anneal) perform better in reliability testing than do dry oxides. As shown in Fig. 3, the charge-to-breakdown (Qbd) of these wet oxides is greatly improved over dry (O_2 oxides) when subjected to DRAM processing.

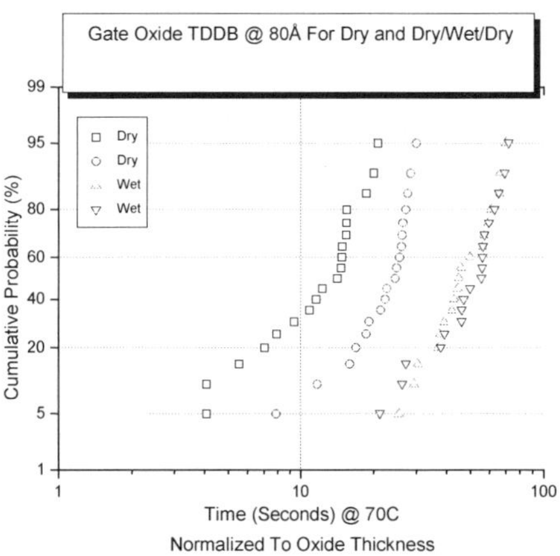

Fig.3 Charge-to-breakdown (Qbd) from a constant electric field stress of gate oxides that have gone through a simulated DRAM process flow. Wet oxides (H_2O, blue-green) clearly have a higher Qbd than dry (O_2 only, red-black) oxides.

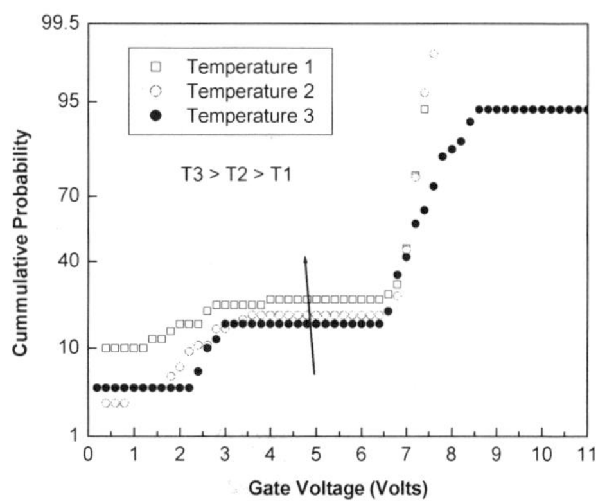

Fig.4 Cumulative probability of breakdown for gate oxides from a DRAM process flow. Decreasing the temperature (T3>T2>T1) at which these oxides were grown is shown to increase the defect density for the early defect failures of the oxides.

However, it is also true that higher temperatures of growth also may provide an advantage to gate oxides. As shown in Fig. 4, decreasing oxide growth temperature increases the defect tail of the intrinsic defect strength of the oxide subjected to a ramped voltage stress. Thus, DRAM gate oxide performance drives the gate oxide growth regions to higher temperatures with steam. Obviously, as DRAM technology continues to scale down, the decreasing gate oxide thickness may eventually make a higher temperature, single wafer steam process preferable to that of a lower temperature batch system. The shorter processing time also increases the throughput and therefore lowers the COO.

An ideal single wafer system for DRAM gate oxides should have at least the following characteristics:

- Quartz-lined chamber for steam processing as well as possible use of chlorine source for metals gettering.
- A small volume for rapid ambient changes such that a substantial portion of the growth is in the ideal regime.
- Load lock and ambient control for staging with pre-gate oxide growth cleans.
- Multi-chambers (>2) on a small platform for improved throughput and foot print (COO).

Another process necessity that may drive DRAMs to single wafer gate oxides is the need for surface P-channel devices. For boron penetration, conventional anneals in nitric or nitrous have been shown to be effective in stopping boron. However, relative to a logic process, the higher thermal budget of a DRAM requires a higher amount of nitrogen at the oxide-silicon interface. This high amount of nitrogen will lead to device transconductance loss and

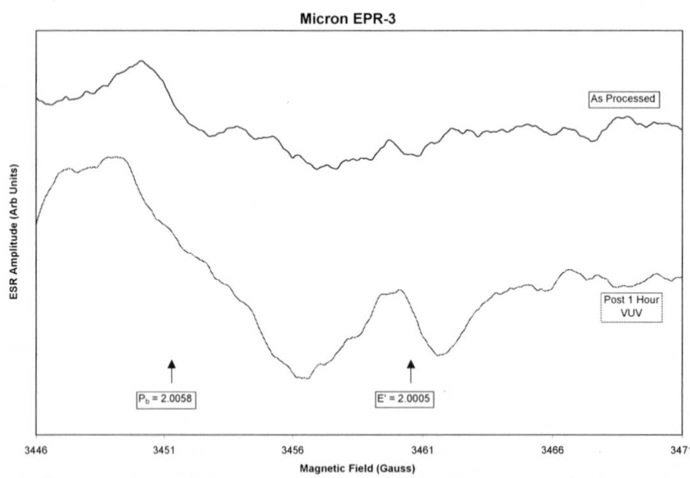

Fig.5 Electron spin resonace spectra from a gate oxide on silicon, as grown (blue) and post-UV light stress (red). The E` (oxygen vacancies) and Pb (interface states) are clearly shown in these 40Å oxides.

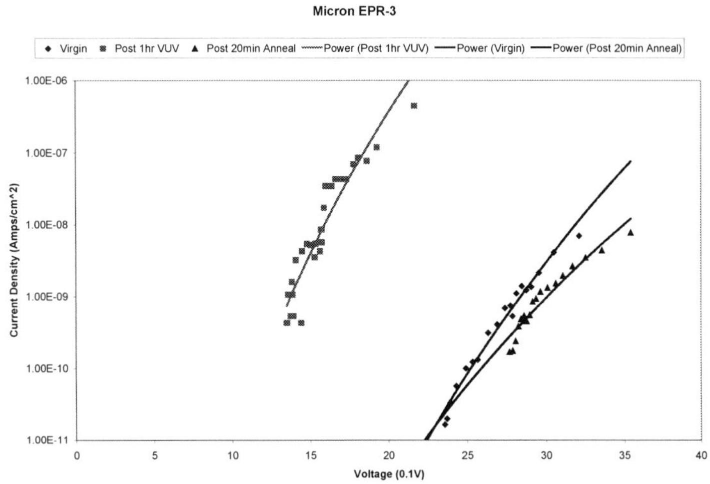

Fig.6 Measurement of corona ion leakage through the gate oxide, virgin (diamonds), post 1-hour UV stress (square), and post-atmospheric anneal at 250°C (triangle).

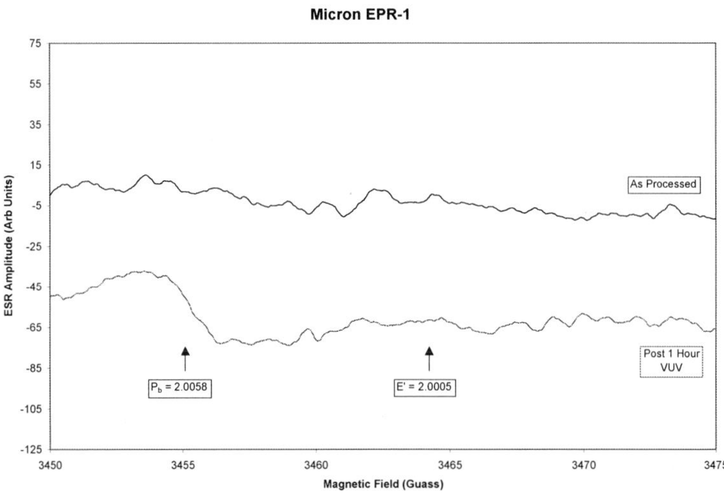

Fig.7 Electron spin resonace spectra from a gate oxide with nitric oxide anneal on silicon as grown (blue) and post-UV light stress (red). The E` (oxygen vacancies are reduced relative to the standard gate oxide.

reliability issues as shown in the literature [2–5]. As shown in Fig. 5, a 40Å oxide grown by conventional means shows the typical interface states (Pb centers) and oxide vacancies (E' centers) as analyzed by electron spin resonance [6–8]. These oxides were then subjected to a UV-light stress (carrier injection). The relative number of interface states, as well as the relative number of oxygen vacancies in the film, is shown to increase post-UV stress. This increase in oxide defects is reflected in the increase in Corona leakage through the films as shown in Fig. 6. As shown, however, the leakage degradation can be recovered through an atmospheric anneal at 250ºC.

For nitric annealed wafers, though, the relative number of E` and Pb centers are lower prior to stress and do not increase as much as the standard oxide post-UV stress, as shown in Fig. 7 and Fig. 8. This is consistent with previous published estimations of interface traps/states [9]. However, in the case of the nitric annealed wafers, the atmospheric anneal does not recover fully the leakage current. Thus, some unpassivated leakage paths exist post-stressing in these films. This may ultimately lead to reliability issues with the DRAM process flow. This issue compounds the issue of transconductance loss for DRAM applications.

As one potential alternative process, composite gate dielectrics of oxide and higher K material may be potential candidates for DRAM devices. Although composite nitride-oxide gates may be precluded from logic applications due to the high effective oxide thickness (EOT), these may be useful for the relatively thick DRAM gate dielectrics [10–11]. Many parameters have made single wafer processing the choice for these composite gates. The most prevalent are the choice of nitride growth precursor for high nucleation on oxide and the need for pre- and/or post-nitride treatments for reduced dielectric charges in the final composite film.

5.Advanced Oxidation

For standard wordlines with tungsten silicide gates, it is known that oxidation of the poly

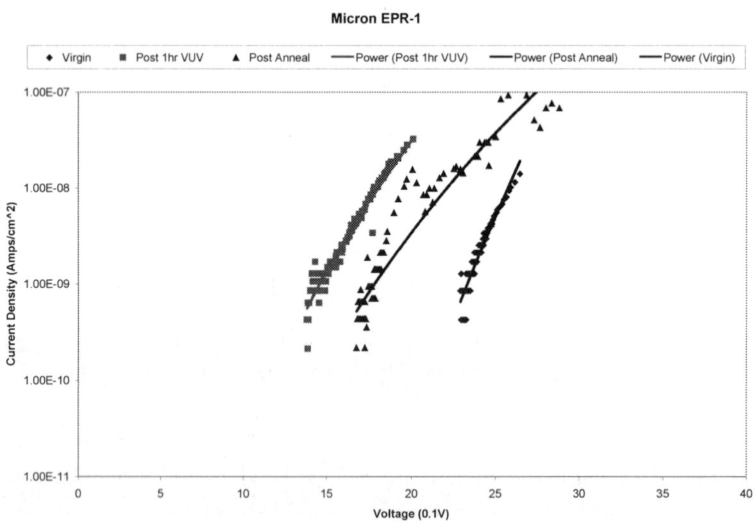

Fig.8 Measurement of corona ion leakage through the gate oxide with nitric oxide anneal, virgin (diamonds), post-1 hour UVstress (square) and post-atmospheric anneal at 250°C (triangle).

gate sidewall and gate oxide edge post-patterning with dry etch is needed for improved reliability of the device. As these device widths shrink, the amount of oxidation under the gate becomes critical. The process demands a minimal amount of oxidation under the gate while still oxidizing the sidewall of the poly silicon wordline. To achieve this goal, many logic companies have turned to single wafer systems [12]. These systems allow for higher oxidation temperatures with a minimal thermal budget. The higher oxidation temperatures therefore allow for higher poly oxidation rates (reaction rates) relative to the oxide thickening (diffusion limited). DRAM devices may ultimately find such advantages with single wafer solutions.

Device speed improvements may ultimately push DRAM devices to move to advanced gate stacks such as TiSix, CoSix, or tungsten. Equipment manufacturers are already exploring selective oxidation of silicon (relative to tungsten) to use in the oxidation of the wordline and gate oxide [12]. As shown in Fig. 9, selective oxidation may be a very successful single wafer application. The need for strict ambient control in isolating wafers from free oxygen as well as reportedly nitrogen would favor the use of a smaller, single wafer chamber over the large volume batch systems.

The advantage of high temperature oxidation in a single wafer tool may ultimately be needed for STI oxidation liners. The potential for a more uniform oxidation of the trench sidewall relative to the corner rounding of the trench may ultimately show a large process advantage for future DRAM generations.

6.Ultra-Shallow Junctions

To date, the DRAM thermal processing budget post-critical transistor implants (for DRAMs with non-trench cells) is still too high to benefit from a spike anneal of the boron implant [13–16]. To demonstrate this, BF2 was implanted into silicon through a thin oxide surface and processed as follows: one wafer was spike annealed, one wafer was run through a simulated DRAM process flow, and one wafer was spike annealed immediately post-implant and then run concurrently through the DRAM process flow. As shown in Fig. 10, the SIMs

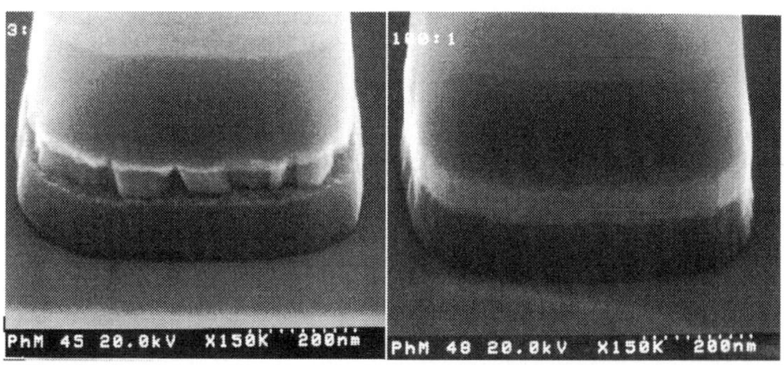

Fig.9 Comparison of SEM cross section after RTP oxidation in (a) a non-selective regime (950°C 30% steam) and (b) selective regime (950°C 1% steam) for test structure.

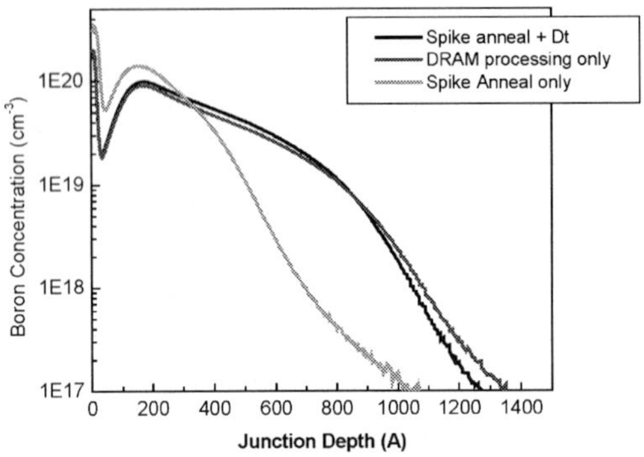

Fig.10 SIMs profile of boron from a BF2 implant into silicon, post-spike anneal only (green), post-DRAM processing (red), and spike anneal with DRAM processing (black).

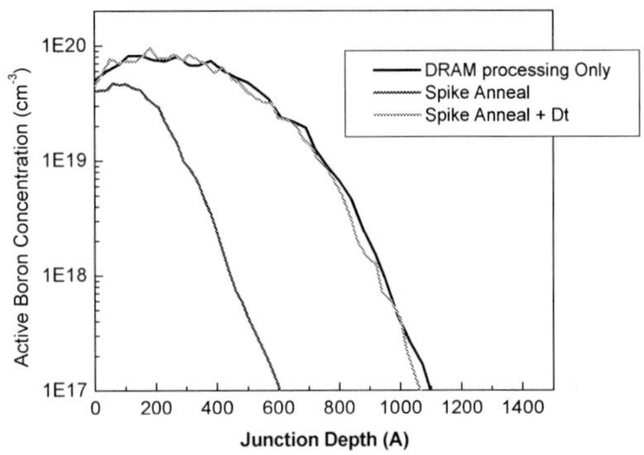

Fig.11 SRP profile of boron from a BF2 implant into silicon, spike anneal only (red), DRAM processing (black), and spike anneal with DRAM processing (green).

profile of the boron in the wafer showed little if any difference between the DRAM process and the spike anneal + DRAM process. The SRP profiles of Fig. 11 also show that the active carrier concentration profiles are very similar. Previous work has shown that the P-channel subthreshold voltage of a DRAM device is a good measure of thermal budget. Transistor devices show that the subthreshold voltage for two different spike anneals immediately post-transistor implants do not improve the DRAM device performance (Fig. 12). The devices are actually made more aggressive by the additional thermal budget as shown by the increase (more positive) subthreshold voltages.

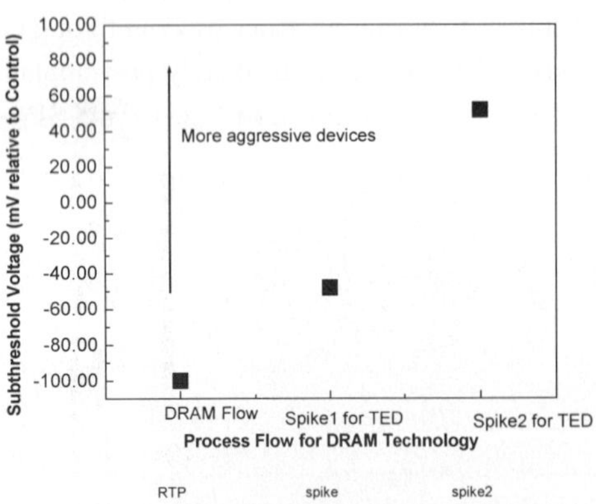

Fig.12 Sub-threshold voltage of DRAM device with a spike anneal of critical implant prior to DRAM processing relative to standard DRAM processing only.

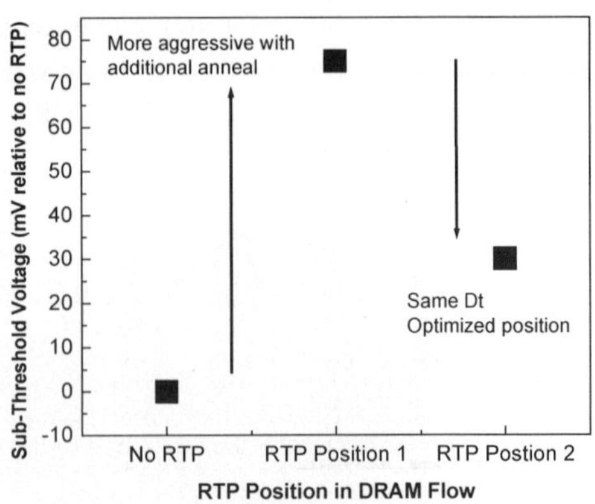

Fig.13 Subthreshold voltage of DRAM device with additional RTP anneal in process flow.

7.RTP Activation Anneals

A rapid thermal anneal can be used as a possible activation for the doped polysilicon or active areas in a DRAM device. As shown in Fig. 13, placement of the RTP activation anneal can be critical for the DRAM transistors. For the same thermal budget, the RTP can have drastically different effects on the transistor leakage characteristics (important for DRAM refresh) as measured by the change in the subthreshold voltage.

One potential application of spike anneals would be to activate dopants with reduced overall thermal budget. As shown in Fig. 14, doped poly is activated with a spike anneal with the resulting Rs measured relative to a more conventional RTP anneal. Again, the placement of the RTP relative to the other thermal processing is important. In Fig. 15, the resulting crystal structure from the initial crystallization anneal affects the final Rs in addition to the final activation anneal. In Fig. 16, however, it is shown that increased cell plate activation due to RTP activation reduces the poly cell plate depletion while improving the capacitance. These gains in speed and capacitance are not limitless [17]. As explored previously, increased thermal budget can increase the number of DRAM manufacturing defects even as overall speed of the devices increases as shown in Fig. 17.

8.LPCVD Processes

In general, the advantage of an LPCVD single wafer technology over the batch process is not clearly established for FEOL DRAM production. The COO of most single wafer processes, such as polysilicon, nitride, LPCVD oxide, and so on, are very large relative to a batch process. The cost of ownership is not only the cost of single wafer tools and the number of tools needed relative to batch technologies, but it is also the cost of down time; the need for cleans, chamber pre-treatments, and equipment maintenance; and consumable costs.

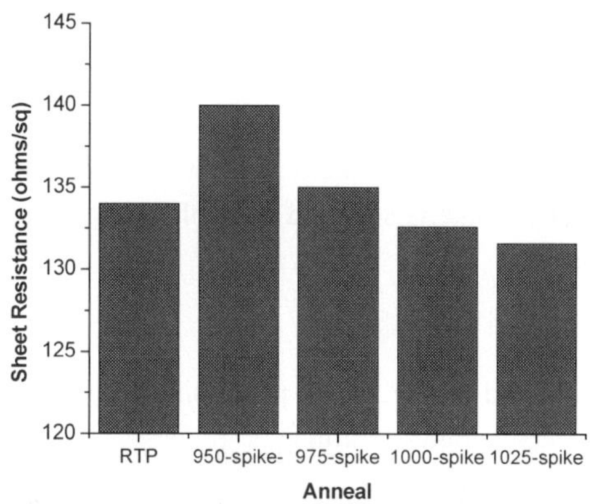

Fig.14 Sheet resistance of doped poly silicon.Comparison of standard

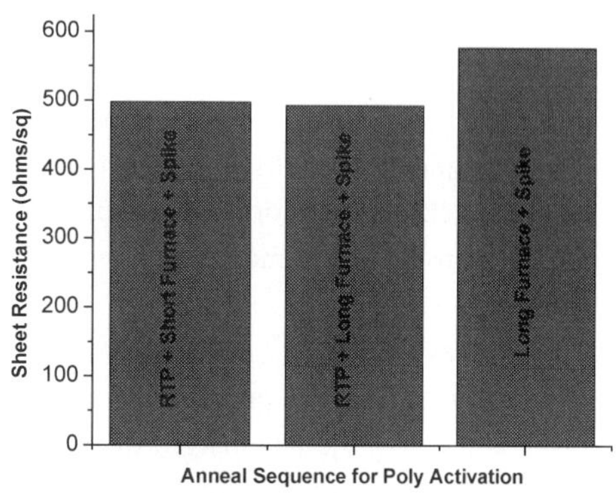

Fig.15 Sheet resistance of doped polysilicon relative to position of RTP anneal in DRAM process flow.

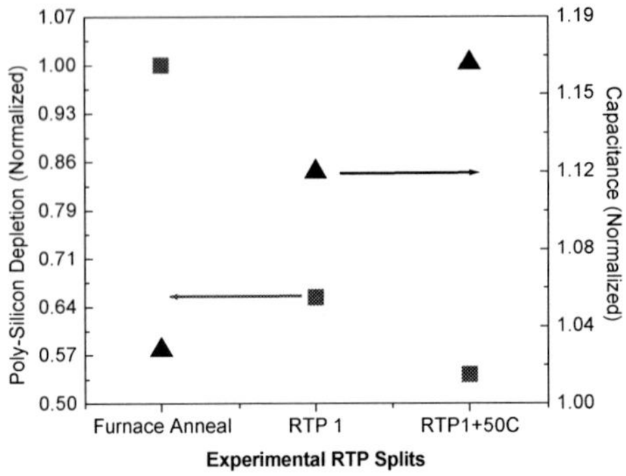

Fig.16 Effect of increased activation of polysilicon cell plates on capacitance of DRAM cell capacitor.

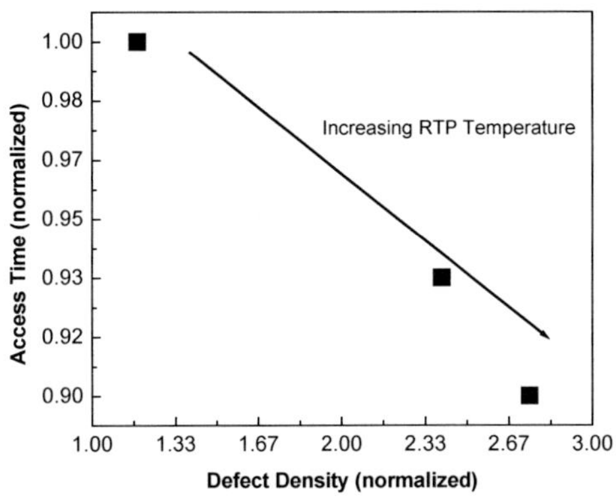

Fig.17 Effect of addtional RTP anneal on DRAM devices. Increasing activation anneal increases speed and DRAM defect density.

Established single wafer technologies for FEOL are ones that have deposition chemistries that are not conducive to batch processing, such as silicide deposition for the wordline. However, other processes such as selective rough polysilicon and ultra-thin cell dielectrics that may benefit from stringent ambient control may ultimately benefit from single wafer processing.

For LPCVD batch processes, reducing the deposition parameters of temperature and time is only now becoming important for DRAM production. While much less critical to the overall DRAM thermal budget than oxidation and RTP reflow processes, single wafer tools may at some future point provide some advantage for DRAMs. As shown in Fig. 18, the thermal budget saved by a fast ramp LPCVD process over a standard furnace already noticeably affects advanced DRAM devices.

The focus for single wafer tools for these established LPCVD batch processes should, however, be on advanced chemistries that fundamentally alter the property of the material. For example, to benefit DRAMs that rely on a hard mask pattern for dry etch control, advanced chemistries for silicon nitride or alternative materials should be explored on single wafer to provide more margin for dry etch chemistries.

9.Summary

Advanced DRAM processing will migrate toward single wafer processing only when process requires the advantages of single wafer technologies. Ultimately, processes such as the oxidations of the wordline poly or active area for gate dielectric formation may become single wafer processes due to the need for ambient control in future DRAM devices. Other processes that are post-critical DRAM transistor implants will benefit eventually from the lower thermal budget of the single wafer processing relative to batch systems. Most processes

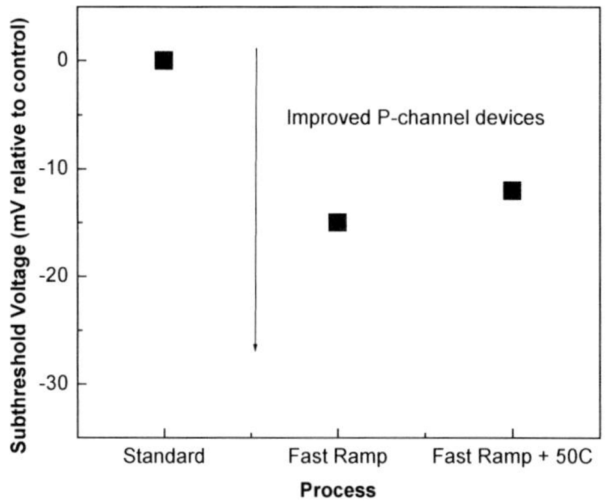

Fig.18 Improvement in DRAM transistors due to fast ramp furnace for process post-critical transistor implants.

to date, however, will need to show some distinct process advantage for a single wafer technology to overcome the large COO associated with a single wafer technology relative to a batch system.

References

[1] Ronald A. Weimer, Proc. Electrochemical Society Meeting, ULSI Process Integration, p. 59, Honolulu, HI (1999).

[2] T. Hori, *IEEE Trans. Electron Devices*, **37**, p. 2058 (1990).

[3] H. S. Momose, T. Morimoto, K. Yamabe, and H. Iwai, *IEDM Tech. Dig.*, p. 65 (1990).

[4] M. Khare, X. W. Wang, and T. P. Ma, *IEEE Electron Device Lett.*, **20** (1), p. 57 (1999).

[5] Mariko Takayanagi-Takagi and Yoshiaki Toyoshima, *IEDM Tech. Dig.* (1998).

[6] P. M. Lenahan, J. J. Mele, J. P. Campbell, A. Y. Kang, R. K. Lowry, D. Woodbury, S. T. Liu, R. Weimer, P. M. Lenahan, and J. F. Conley, Jr., *J. Vac. Sci. Technol.* **B16** (4), Jul/Aug, p. 2134 (1998).

[7] W. L. Warren, M. R. Shaneyfelt, D. M. Fleetwood, J. R. Schwank, P. S. Winokur, and R. A. B. Devine, *IEEE Transactions on Nuclear Science*, **41** (6), p. 1817 (1994).

[8] D. M. Fleetwood and N. S. Saks, *J. Appl. Phys.* **79** (3), p. 1583 (1996).

[9] G. Lucovsky, *IBM J. Res. Develop.*, **43** (3), p. 301 (1999).

[10] Anri Makajima, Takashi Yoshimoto, Toshiro Kidera, Katsumori Obata, Shin Yokoyama, Hideo Sunami, and Masataka Hirose, *Applied Physics Letters*, **77** (18), p. 2855 (2000).

[11] John H. Das, Carl Powell, Vidula Kirtikar, April D. Daniel, Ron Weimer, and Sing Pin Tay, IEEE 8th International Conf. on Adv. Thermal Proc. of Semiconductors RTP2000, p. 154, Gathersberg, MD, Sept 20–22, (2000).

[12] Richard B. Fair, *IEEE Transactions on Electron Devices*, **37** (10), p. 2237 (1990).

[13] P. M. Fahey, P. B. Griffin, and J. D. Plummer, Rev. Mod. Phys., **61** (2), p. 289 (1989).

[14] K. S. Jones and G. A. Rozgonyi, *Rapid Thermal Processing Science and Technology*, Ch. 5, Academic Press, NY (1993).

[15] J. Hebb and K. F. Jensen, 4th Int. Conf. on Advanced Thermal Processing of Semiconductors RTP '96, edited by R. B. Fair, M. L. Green, B. Lojek, and R. P. S. Thakur, Boise, ID (1996).

[16] R. A. Weimer, D. Ratakonda, C. Powell, M. Nuttal, R. P. S. Thakur, P. J. Timans, and N. Shah, 5th Int. Conf. on Rapid Thermal Processing of Semiconductors RTP '97, p. 91, New Orleans, LA (1997).

Ultra-high vacuum rapid thermal chemical vapor deposition for formation of TiN as barrier metals

S. Naito[1], M. Okada[1], O. Nakatsuka[2], T. Okuhara[4], A. Sakai[1], S. Zaima[3], and Y. Yasuda[1]

[1]Department of Crystalline Materials Science, Graduate School of Engineering, Nagoya University
[2]Center for Integrated Research in Science and Engineering, Nagoya University
[3]Center for Cooperative Research in Advanced Science and Technology, Nagoya University
Furo-cho, Chikusa-ku, Nagoya, 464-8603, Japan
TEL: +81-52-789-3923, FAX: +81-52-789-5592,
e-mail: nakatuka@alice.xtal.nagoya-u.ac.jp
[4]Toyoko Kagaku Co., Ltd.
1280 Nakamaruko, Nakahara-ku, Kawasaki, 211-8502, Japan
TEL: +81-44-435-5866, FAX: +81-44-434-6303

1.Introduction

In recent years, there are increasing demands for future ULSI devices such as decreased size of ULSI circuits and wide variety of functions on a device chip. It is necessary to develop ultra thin film growth technologies which offer several critical requirements; high controllability of the film thickness, high controllability of the grain orientation, low contaminants in the films, good step-coverage, uniformity of the film thickness, and flexibility of the selection of materials used for sorts and impurities. Ultra-high vacuum rapid thermal chemical vapor deposition (UHV-RTCVD) is expected to be one of powerful methods which meet these requirements and has several advantages. First, the ultra-high vacuum system provides low partial pressure of H_2O and O_2 during growth, which avoids hydrolysis of reactive source gases and suppresses unintentional oxidation of deposited films and substrates. Second, the rapid thermal process (RTP) [1-2] with rapid heating and quenching provides accurate control of reaction source gases so that the incorporation of contaminants due to by-product of the reaction are greatly suppressed. Therefore it is possible to deposit thin films with excellent quality by the UHV-RTCVD.

In this work, we have developed an UHV-RTCVD system and grown TiN films using the system. CVD-TiN films grown using $TiCl_4$ and NH_3 as source materials are attractive candidates used for the multilevel metallization in the ULSI devices. Especially, the TiN film as a diffusion barrier metal for metallized Cu has been required to be thin enough and have low resistivity, low contaminants in the films, and good step-coverage. We have applied the UHV-RTCVD method to the TiN film growth and investigated structural and electrical properties.

2.Experimental

UHV-RTCVD system

Figures 1 shows a schematic illustration of an UHV-RTCVD system which we developed. It can be equipped with three growth chambers, each of which is bakeble to attain ultra-high vacuum as shown in Fig. 1(a). Each growth chamber is connected through transfer chamber and wafers are transferred from a load-lock chamber into the growth chambers by a robot arm. The growth chambers can be evacuated with alternative pumping systems; two turbo molecular pumps with a rotary pump or turbo booster pump with a dry pump.

Figure 1(b) shows a detailed structure of the TiN growth chamber which is mainly made with quartz and placed between infrared halogen lamp heaters for rapid heating. Five circle heaters with 900–4000W can be controlled separately for ensuring temperature uniformity of

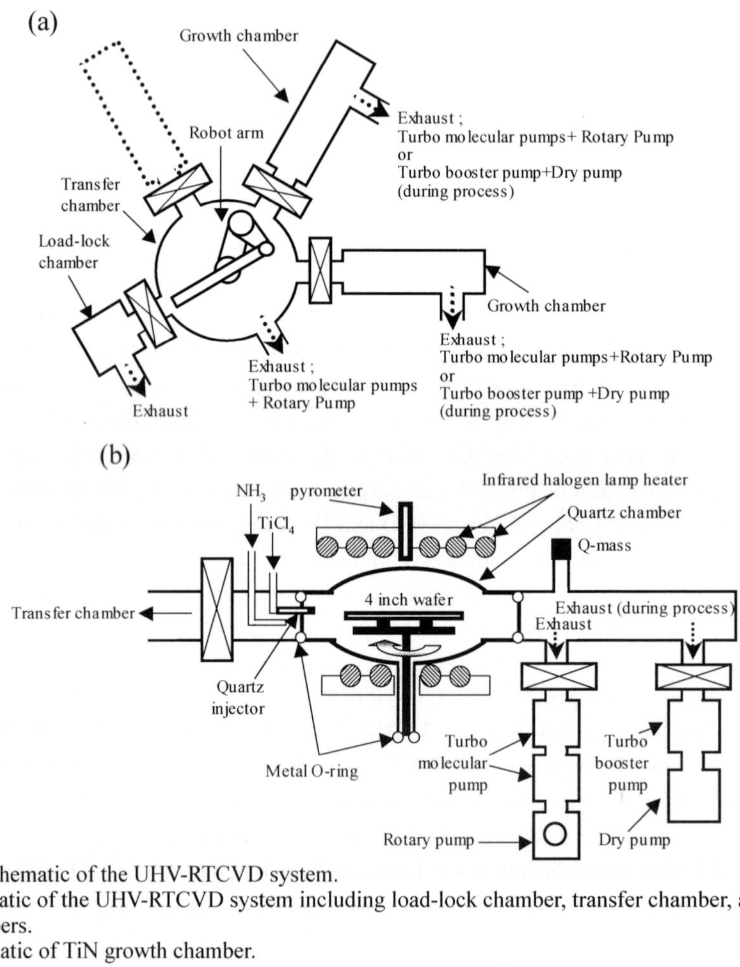

Fig.1 Schematic of the UHV-RTCVD system.
(a) Schematic of the UHV-RTCVD system including load-lock chamber, transfer chamber, and growth chambers.
(b) Schematic of TiN growth chamber.

Figure 1(b) shows a detailed structure of the TiN growth chamber which is mainly made with quartz and placed between infrared halogen lamp heaters for rapid heating. Five circle heaters with 900–4000W can be controlled separately for ensuring temperature uniformity of the substrate surface. The substrate table can be rotated to provide the uniformity of film's characteristics in a 4-inch wafer scale. The substrate temperature is measured by a pyrometer sensitive to long wavelength infrared radiation in order to permit the measurement of temperature ranging from 400°C to 1100°C. The pyrometer is set up to view the substrate surface through the quartz chamber. During film growth, the growth chamber is evacuated by a turbo booster pump and a dry pump. This realizes an oil-free environment and thus prevents unintentional carbon contamination in the growth chamber. $TiCl_4$ and NH_3 as source gases to form TiN films are introduced into the growth chamber through separate lines and gas injectors so that reaction takes place near the hot substrate surface. The $TiCl_4$ gas injector is made with quartz which is transparent for the infrared radiation in order to prevent thermal decomposition of $TiCl_4$ by inside the gas injector.

TiN film growth

SiO_2 layers with a thickness of 80 nm formed by thermal oxidation on Si (100) wafers were used as substrates. The base pressure was less than $1x10^{-8}$ Torr before deposition. The deposition was performed at substrate temperatures ranging from 450°C to 650°C with a $TiCl_4$ gas flow of 3 sccm and a NH_3 gas flow of 20 sccm with a N_2 carrier gas. The total pressure during the deposition was 10 mTorr and the partial pressures of $TiCl_4$ and NH_3 were, respectively, 0.73 mTorr and 4.9 mTorr which were calculated by the method of Buiting [3].

The thickness of the deposited films was measured by a stylus surface profiler. The crystalline structure of the TiN films was analyzed by X-ray diffraction (XRD) and transmission electron microscope (TEM). The surface morphology was observed by atomic force microscope (AFM). A four points probe method was used for the measurement of the resistivity. The contaminant in the film was measured by Auger electron spectroscopy (AES).

3.Results and Discussion

Figure 2 shows deposition time dependence of the thickness of TiN films deposited at various substrate temperatures. The growth rates were determined from the gradient of each plot, which were 3.6 nm/min, 8.8 nm/min, and 8.5 nm/min for the temperatures of 450°C, 550°C, and 650°C, respectively. The growth rate increases with the increase in the substrate temperature and appears to be constant at the temperatures over 550°C, suggesting that the growth is limited by the gas flow rates of $TiCl_4$ and NH_3. These values of the growth rates are lower than those in low pressure CVD (LPCVD) reported previously [3-6], which comes from the lower partial pressure in our RTCVD compared with that of the LPCVD, typically about 100 mTorr.

Figure 3 shows low incident angle XRD profiles taken from the films grown at 450°C and 550°C for various deposition times. The peaks related with TiN (111), (200) and (220) were clearly observed in all profiles of 550°C samples, which indicates that polycrystalline phases with a NaCl type structure are formed. On the other hand, in the case of 450°C, peaks related to TiN are hardly observed not only for the 4-min-deposited film but also for the 5-min-deposited film whose thickness is nearly equal to that of the sample grown at 550°C for 4 min. This result suggests that a growth temperature above 550°C, at least, is required to obtain crystalline TiN films.

Figure 4(a) shows a transmission electron diffraction pattern taken from a TiN film deposited at 550°C for 5 min. Diffraction spots corresponding to TiN (111), (200), (220), (311), (222), (400), (331), (420), and (422) with a NaCl type structure are observed and this is consistent with the XRD results shown in Fig. 3. A bright-field plan-view TEM image of the same sample is shown in Fig. 4(b). Textures consisting of polycrystalline TiN grains are clearly observed. An average grain size was measured to be about 12 nm in diameter. Compared to the value (9.4 nm) for the 3-min deposited sample measured from its TEM image (not shown), the increase in the grain size in plan-view is smaller than that in the film thickness. This suggests that the film TiN has columnar grain structures. Figure 4(c) is a bright-field cross-sectional TEM image for the same sample as that shown in Fig. 4(b). Polycrystalline TiN films and grain boundaries are clearly observed and the TiN layer was confirmed to have columnar grain structures.

AFM images of TiN films deposited at 550°C for 5 min are shown in Fig. 5(a). An average diameter of the observed dots was measured to be about 20 nm and this value does not agree with that obtained by the plan-view TEM image show in Fig. 4(b). This results suggests that

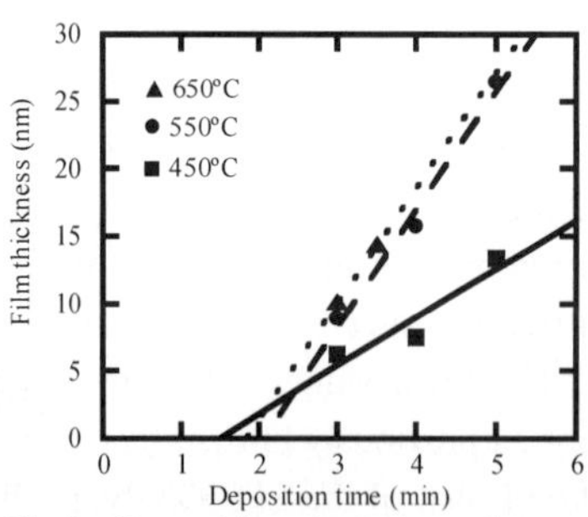

Fig.2 Deposition time dependence of the thickness of TiN films deposited at various temperature by the UHV-RTCVD

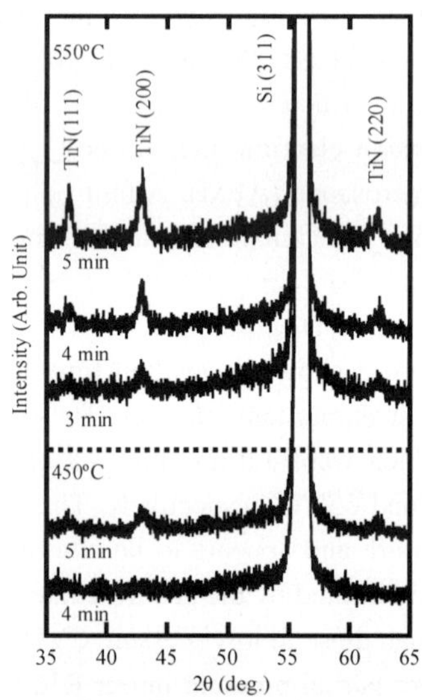

Fig.3 XRD profiles taken from TiN films dep osited at 550°C for 3, 4, and 5 min and deposited at 450°C for 4min and 5min.

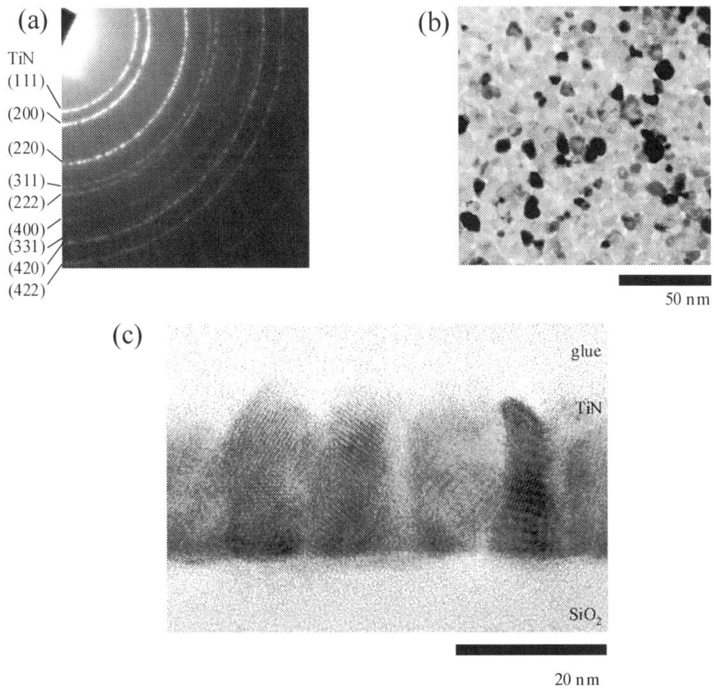

Fig.4 TEM observations for the same sample deposited at 550°C for 5 min.
(a) Transmission electron diffraction pattern.
(b) Bright-field plan-view TEM image.
(c) High resolution cross-sectional TEM image.

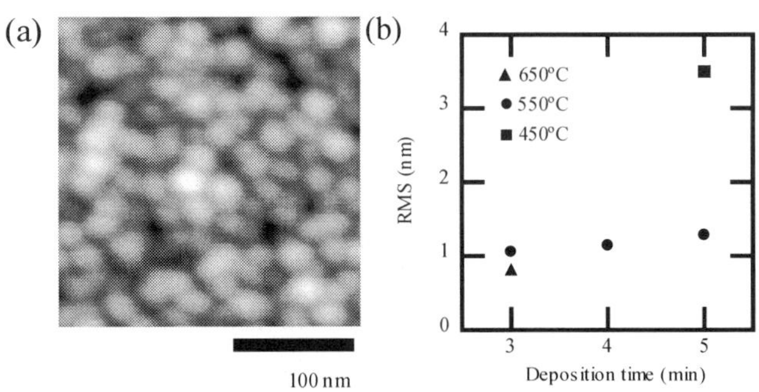

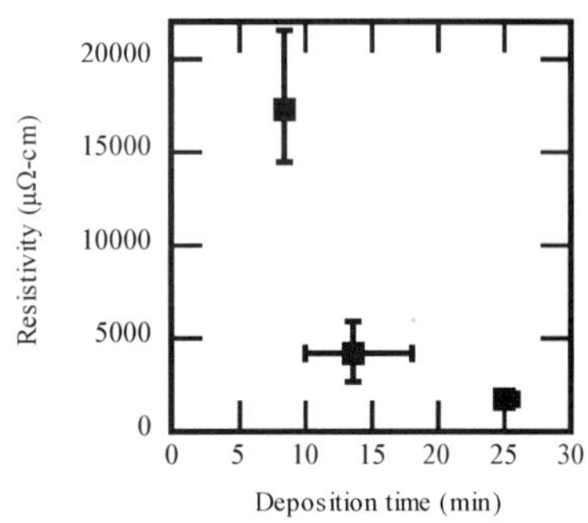

Fig.5 AFM observation for the TiN films.
(a) AFM image of the film deposited at 550°C for 5 min.
(b) RMS value of surface roughness of the TiN films deposited at various temperatures as a function of deposition time.

Fig.6 Film thickness dependence of the resistivity of TiN films deposited at 550°.

The averages values measured five points in range of center to edge in the wafers deposited TiN films at intervals of 5 min are shown.

the AFM image did not reveal individual grains but reflect the surface morphology of coalesced grains. Figure 5(b) shows root mean squares (RMS) values of film surface roughnesses estimated from the AFM observations for samples deposited at 450°C, 550°C, and 650°C. The RMS values of about 1 nm were obtained for 550°C- and 650°C- deposited samples and this value does not change very much with increasing the deposition time. Compared to these, the surface of the sample deposited at 450°C for 5 min was much rougher and the RMS value of 3.5 nm was obtained.

Figure 6 shows the resistivity of TiN films deposited at 550°C as a function of the film thickness. The resistivity decreases with the increase in the film thickness and approaches to the values of TiN films grown by the other $TiCl_4$-based CVD [3-6, 8-10]. The observed higher resistivity for thinner films should be caused by the oxidation of the film when it is exposed to the atmosphere after the growth. It has been reported that TiN film surfaces are oxidize and the TiN-oxide thickness amounts to 6 nm [6,7]. Therefore, the surface of our TiN films might be oxidized and thus the film resistivity increases when decreasing the film thickness.

Figure 7 shows AES spectra of the sample deposited at 550°C for 5 min, which were obtained after *in situ* Ar^+ sputtering. In all spectra, Ti_{LMM}+N_{KLL}, Ti_{LMM}, O_{KLL} peaks are observed. O_{KLL} peak intensity in the as-deposited sample is larger than in the samples after sputtering for 2000 s and 4000 s, which indicates the surface oxidation of TiN films due to exposure to the atmosphere. It is found that the intensities of Si_{LMM} peak appeared at 77 eV and the O_{KLL} peak increase after 6000 s and kinetic energy of the O_{KLL} peak shifts to a lower energy side by about 7.5 eV. This means that the sputtering time longer than 6000s is enough to expose the surface of the substrate SiO_2 layer. Nevertheless, the Cl_{LMM} peak is hardly observed at every stage of the sputtering of the TiN films. On the other hand, we confirmed that Cl_{LMM} peaks appeared in the TiN films deposited at 450°C (not shown). Therefore, the residual chlorine concentration in the film critically depends on the deposition temperature and is less than the AES detection limit in the 550°C-deposited films.

4.Summary

We have developed an UHV-RTCVD system and examined the characteristics of TiN films formed with this system using $TiCl_4$ and NH_3 as source gases. It was found that TiN films had a polycrystalline texture which consisted of nano-meter scale grains having columnar structures. Smooth surface morphology, with an about 1 nm RMS value, was obtained for deposition temperatures about 550°C. The resistivity of the films was higher for thinner films but was comparable to the ordinary values reported previously when the thickness amounted to about 20 nm. The residual chlorine concentration in the film depends on the deposition temperature and the low chlorine concentration was obtained for 550°C-deposited films, which was less than the detection limit of the AES measurement.

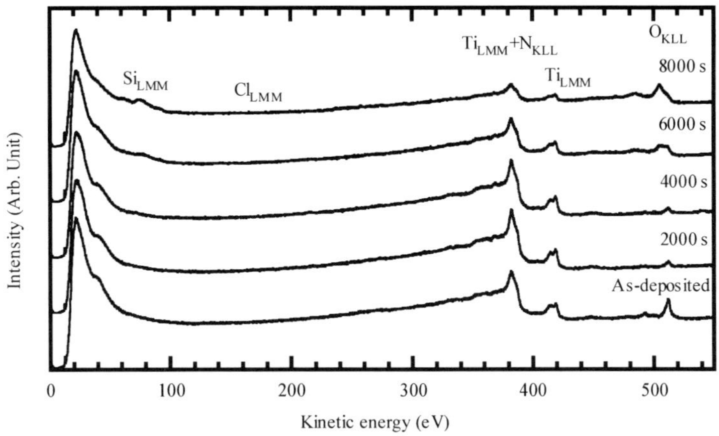

Fig.7 AES survery spectra of the TiN film deposited at 550°C for 5 min. The times on the right-hand side of the figure in dicate the sputtering time with Ar^+

References

[1] J. Niess, CH. Grunwald, R. Strohmaier, and Z. Nényei, J. Electronic Materials **27** (1998) 1286.

[2] Alan Bratschun, J. Electronic Materials **28** (1999) 1329.

[3] M. J. Buiting, A. F. Otterloo, and H. Montree, J. Electrochem. Soc. **138** (1991) 500.

[4] J. B. Price, John O. Borland, and Steve Selbrede, Thin Solid Films **236** (1993) 311.

[5] N. Yokoyama, K. Hinode, and Y. Homma, J. Electrochem. Soc. **138** (1991) 191.

[6] R. I. Hegde, R.W. Fiordalice, E. O.Travis, and P. J. Tobin, J.Vac. Sci. Technol. **B11** (1993) 1287.

[7] A. Sherman, J. Electrochem. Soc. **137** (1990) 1892.

[8] D. H. Kim and B. Y. Kim, Jpn. Appl. J. Phys. **38** (1999) L461.

[9] R. Leutenecker, B. Fröschle, U. Cao–Minh, and P. Ramm, Thin Solid Films **270** (1995) 621.

[10] H. Hamamura, R. Yamamoto, K. Takahiro, S. Nagata, S. Yamaguchi, H. Komiyama, and Y. Shimogaki, in *Proc. Advanced Metallization 1998* (MRS, Warrendale, PA, 1998) pp. 345-349.

Implementations of Rapid Thermal Processes in Polysilicon TFT Fabrication

Miltiadis K. Hatalis[1] and Apostolos T. Voutsas[2]

[1]Dept. of Electrical and Computer Engineering, Display Research Laboratory, Lehigh University, Bethlehem, PA 18015, USA
Phone:+1-610-758-3944 Fax:+1-610-758-6248 E-mail: MKH1@lehigh.edu
[2] Sharp Laboratories of America, LCD Process Technology Department, Camas, WA 98607, USA
Phone:+1-360-834-8646 Fax:+1-360-817-8500 E-mail: avoutsas@sharplabs.com

1.Introduction

Thin film transistor (TFT) technologies have enabled the realization of active matrix liquid crystal flat panel displays, which have contributed to the widespread use of laptop computers. Currently, emphasis is being placed in the replacement of the cathode ray tube (CRT) displays with flat panel displays in desktop computer systems. Future applications of thin film transistors will include "systems on the panel" in which TFTs, in addition to their use in the system display, could also be utilized in the fabrication of the system processing circuits and memory modules. The realization of such highly integrated systems will require the performance of the thin film transistors to approach that of silicon single crystal CMOS devices. In addition to the TFT high performance requirement, the fabrication of these devices needs to take place at low thermal budgets in order to be compatible with low cost glass substrates. The polycrystalline silicon (polysilicon) TFT technology is being currently actively pursued for such "system on panel" applications [1]. A critical feature of the polysilicon TFT technology is the use of Rapid Thermal Processing in order to achieve both the high performance and the low thermal budget requirements. The fabrication of polysilicon TFTs requires three different high temperature process steps, which in order of implementation, are: a) dehydrogenation of amorphous silicon, b) crystallization of amorphous silicon and c) activation of ion-implanted dopants. This paper focuses on the crystallization of amorphous silicon for the formation of the polysilicon TFT active layer and presents recent advances on the Excimer Laser Annealing (ELA) and xenon-lamp based RTP.

2.Amorphous Silicon Crystallization Processes.

Excimer Laser Crystallization

Excimer laser crystallization relies on the strong absorption of UV light by Si and the corresponding coupling of photon energy to the Si network. This results in the very rapid heating of the Si material at or above its melting point. Since the annealing process involves melting and recrystallization, significantly improved microstructure can be obtained vis-à-vis solid-phase crystallization methods. The final microstructure is very sensitive to the conditions of the irradiation. In fact, the sensitivity and instability of this process has long inhibited the wide application of excimer laser annealing as a mass-production worthy crystallization method. Research in this area dates back to 10+ years and has led to a much better understanding of, and associated advances on, this process.

The excimer laser of choice for Si crystallization is XeCl (308nm). At this wavelength a-Si absorbs very strongly ($\alpha \sim 10^6 cm^{-1}$), whereas the cost and lifetime of the associated optics are suitable for mass-production operations. Typical excimer lasers operate in a pulse mode, at frequencies around 300Hz. The laser pulse is typically very short, with duration in the range of 20-30nsec. On the laser equipment side one significant issue has been the pulse-to-pulse stability. This issue emanates from the need to operate in a highly unstable regime, during crystallization, to obtain the best poly-Si microstructure [2]. Hence, even small variations in the laser fluence that reaches the sample at each pulse can result in significant differences in the film's microstructure (figure 1). As a result, a compromise between material quality and process stability seems necessary to overcome this problem.

Over the past few years, laser equipment makers have made significant efforts to improve the process stability by gradually improving the pulse-to-pulse stability of their excimer lasers. Surprisingly enough, early excimer lasers, based on corona-discharge method, demonstrated good stability characteristics which, however, were compromised, by moving into other pre-ionization techniques, to benefit power output. The one exception to this case was the high-energy (15J) laser commercialized by SOPRA (France) that offers a 3% peak-to-peak stability, due to a stable X-ray pre-ionization method [3]. Over the past 2 years, the market leader in TFT excimer lasers (Lambda Physik) has also introduced mass production models with 3σ peak-to-peak variation of better than 9% (and reportedly better than 6%) [4], [5].

These improvements in the laser equipment characteristics are certainly beneficial as far as the process window but, nonetheless, cannot by themselves overcome the sensitivity of laser crystallization to process variations and, furthermore, guarantee the development of

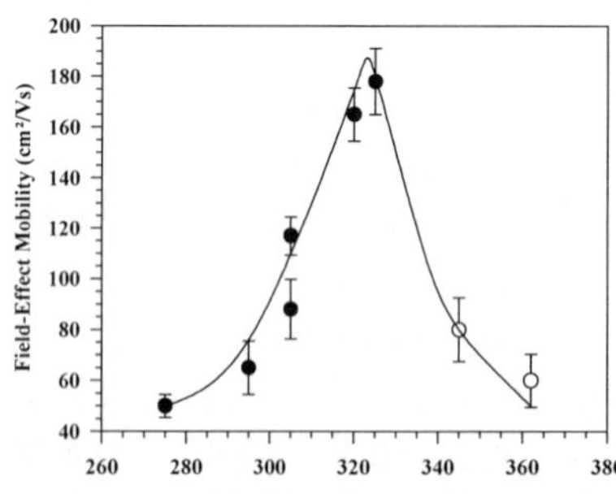

Figure.1 TFT Mobility vs. Laser Fluence for standard laser annealing process

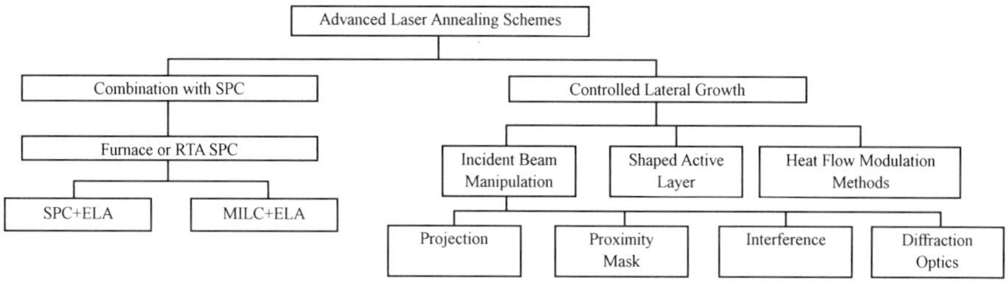

Figure.2 Classification of advanced laser annealing techniques.

poly-Si microstructures with crystal quality compatible with future generations of product (display) requirements. To that end, a whole new class of laser-based crystallization schemes has been developed with various degrees of success as far as flexibility and potential for commercialization [6-8]. Figure 2 shows a "family tree" of such techniques and a suggested classification scheme. There are at least two common denominators among such schemes: (1) Schemes that use laser annealing in conjunction with another crystallization technique (2) Schemes that invoke controlled lateral growth using laser annealing. Some of the techniques shown in figure 2 are briefly described here. Combination of Solid Phase Crystallization (SPC) with excimer laser annealing has been shown to yield improved material quality with a slightly improved process window [9]. A variant of SPC technique, so-called Metal-Induced Lateral Crystallization (MILC) has been quite effective in improving the microstructure of solid-phase crystallized films [10]. When this technique is combined with ELA an even greater improvement of the properties of the resulting poly-Si microstructure has been observed [11]. It should be noted that the initial, SPC crystallization step can be carried either by furnace annealing or rapid thermal annealing (RTA). Thus, the possibility of combining RTA and ELA processes to obtain high quality poly-Si precursor is possible. This presents an excellent opportunity for RTA process to make its way to the poly-Si TFT fabrication flow.

A different set of techniques rely on the concept of controlled lateral growth (CLG) that has been coined by Im's group at Columbia University [8], [12]. Looking at the TFT-mobility-vs.-laser-fluence curve of figure 1 it is clear that the material characteristics are optimized within a very narrow fluence range, below and above of which the material shows rather poor performance. Hence, the idea of controlled lateral growth is to use this optimal fluence range in a manner that is consistent with uniform and reproducible material characteristics. Unlike the traditional approach to laser annealing that tries to avoid at all costs complete melting of the film, the family of CLG schemes aims at complete melting of the film, however, within a very narrow (and controlled) region. The width of such region is typically in the order of a few microns (i.e. 2-5μm). Hence, the incident laser beam (typically having a widthof many 100s of microns) has to somehow be transformed to such fine "beamlets". The transformation process (or "shaping") can be achieved by a variety of masking methods, (i.e. [12]). Other approaches to achieve CLG include pre-patterning of the active layer [13] and/or

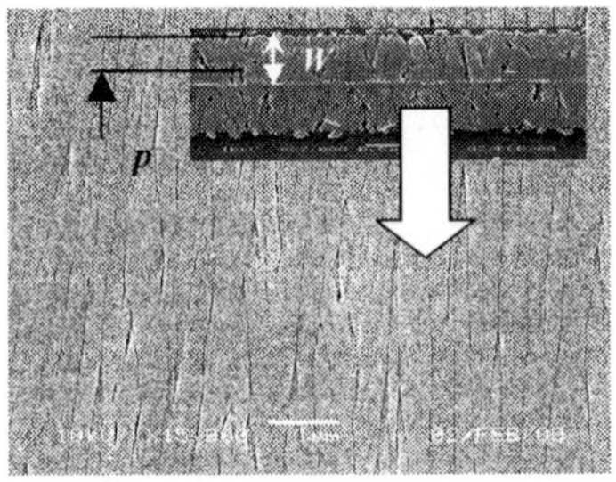

Figure.3 Poly-Si microstructure after multiple translations of the initial seed under the laser beamlet (inset: original seed, i.e. poly-Si microstructure after the first laser irradiation).

a variety of concepts, which aim to alter the heat diffusion patterns during and after the irradiation process [14], [15]. Manipulation of the incident beam through masking (primarily by projection) presents the most flexible and efficient approach to CLG. Such schemes are based on a simple principle, very similar to ZMR: first, a localized seed of laterally grown poly-Si material is generated and then the seed is "dragged" laterally by precise and highly controllable positioning of the substrate under the shaped, laser beamlet. Figure 3 shows an example of the resulting poly-Si microstructure based on this technique (coined Sequential Lateral Solidification, or SLS [16]). The inset in the figure shows the microstructure of the original seed (i.e. after the first irradiation). The direction of motion of the substrate under the shaped beamlet is indicated by the arrow. As long as the substrate moves with a pitch less or equal to the length "p" the resulting microstructure resembles that of figure 3.

Application of standard laser annealing crystallization process results in a poly-Si microstructure that resembles that shown in figure 4(a). It consists of relatively small grains (0.1-0.3μm), having approximately an equiaxial cross-section, with random orientations. Efforts to enlarge the grain size, in order to improve the film's electrical properties, have been based on lengthening the melt duration (i.e. the average time the Si film stays in the molten state during crystallization). Figure 4(b) shows this relationship.

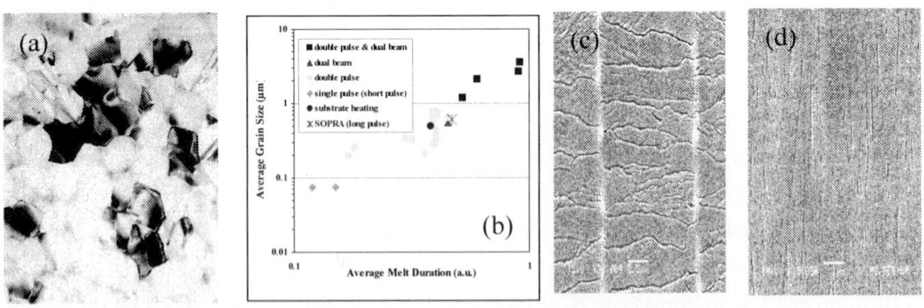

Figure.4 (a) Typical microstructure of standard laser-crystallized poly-Si film. (b) Correlation of average grain size and melt duration. (c) and (d) Microstructure of laser crystallized poly-Si films by advanced process.

Table 1:TFT performance characteristics versus type of laser annealing process (poly-Si film thickness ~1000Å)

Process	Style	Type	TFT Mobility (cm^2/Vs)		Uniformity	References
			Channel //	Channel ⊥		
Std. ELA			150	150	30%	
Adv. ELA	Projection	*Directional (p<w)*	370 340 360	150 140 160	10% 7.5% -	[17] [18]
Adv. ELA	Projection	*Periodical (p>w)*	240 220	80 90	5% 3%	[17]

Despite of the successes in enlarging the grain size of standard-processed ELA poly-Si films, the variance in the grain size and the randomness of the location of grain boundaries render these efforts unsuccessful in simultaneously producing very high quality material with excellent uniformity across the crystallized area. In contrast, advanced laser-annealing schemes, based on CLG, have shown significant promise in achieving such breakthroughs. Figures 4(c) and (d) show the microstructure of poly-Si films obtained by SLS-type laser process with a beam pitch ("p") of p>w -*fig. 4c* - and p<w - *fig. 4d* (for definitions of p and w, see figure 3). It is clear that implementation of such advanced schemes allows the precise manipulation of the resulting poly-Si microstructure, which enables the realization of different material qualities, when and where desired. The application of advanced, laser-based crystallization processes in TFT fabrication makes possible the improvement of both performance and uniformity of poly-Si TFTs, to levels not possible before (see table 1). One disadvantage that is currently noted, for these crystallization schemes, is the existence of performance directionality. In other words, TFT characteristics demonstrate a strong dependence on the degree of misorientation between the direction of carrier conduction in the active layer and the direction of lateral growth in the film.

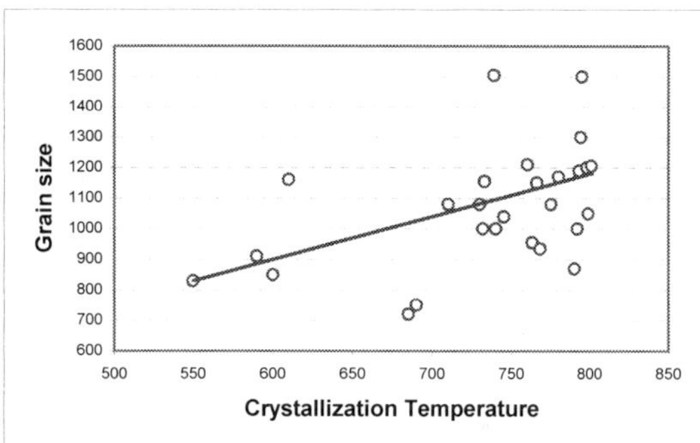

Figure.5 Correlation between crystallization temperature and grain size of RTA crystallized amorphous silicon deposited by PECVD under various deposition conditions

Rapid Thermal Crystallization

An alternative, to the melting and solidification processes that occur during the excimer laser annealing, is the all solid phase crystallization of amorphous silicon. This approach has certain cost advantages over laser annealing and, moreover, lacks the complexities that are associated with the highly non-linear effects in laser crystallization. This process has been studied extensively for furnace crystallized films [19-21]. However, furnace anneal is not suitable for the low cost manufacture of flat panel display on large area glass substrates. Recently, a Rapid Thermal Processing (RTP) system, suitable for large area glass substrates, has been developed by Intevac Corporation [22]. This new RTP approach utilizes two linear arc xenon lamps to rapidly heat and crystallize the silicon film as it is moved underneath the two lamps. RTP promises higher throughput, lower cost and better material uniformity.

The crystallization of amorphous silicon by such RTA process has been recently studied through the use of statistically designed experiments [23]. Figure 5 shows the relationship between the crystallization temperature of the RTA process and the resulting grain size of the crystallized silicon films. All the data shown in Figure 5 are for PECVD deposited amorphous silicon films. The films were deposited under various conditions and this is the reason for the scatter in the data; though there is a trend that indicates that films crystallized at a higher temperature have larger grain size. This is the result of the difference in the nucleation rate (the rate with which new crystallites are formed in the film during the RTA annealing process) of the films deposited under different conditions. In general conditions that favor low nucleation rate during annealing result in larger grain size but are more difficult to crystallize thus requiring higher RTA temperatures. Films deposited by PECVD in general have small grain size as shown in Figure 5. In contrast films deposited by low pressure chemical vapor deposition (LPCVD) had larger grain size. The mobility of the n-channel polysilicon TFTs fabricated in RTA crystallized silicon films was found to depend upon the polysilicon grain size. Typically, higher mobility devices are obtained from crystallized Si films deposited by Low-Pressure CVD (LPCVD), instead of Plasma-Enhanced CVD (PECVD) method. The performance of Si-films deposited by LPCVD can be further classified depending upon the type of precursor gas used for the deposition. Films deposited with disilane gas enable higher mobility than films deposited with silane. This is traced back to the intrinsic properties of the as-deposited films, as a function of the gas precursor, which have been characterized in detail in earlier studies [19], [20]. In brief, disilane gas favors the formation of a-Si films with increased degree of structural disorder that, in turn, yields poly-Si structures with larger grain size upon crystallization. The larger grain size is considered responsible for the increase in TFT mobility. The one drawback, however, is the typically higher crystallization temperature range required to effectively complete the phase transformation [20]. Addition of a "seed" layer has been found to alleviate this requirement, resulting in crystallization at lower annealing temperature. The "seed" layer is essentially supplying nucleation sites and is a thin a-Si layer deposited at conditions that favor nucleation. Once nuclei are formed, they can then consume the a-Si film deposited by disilane gas without the need of nucleation occurring

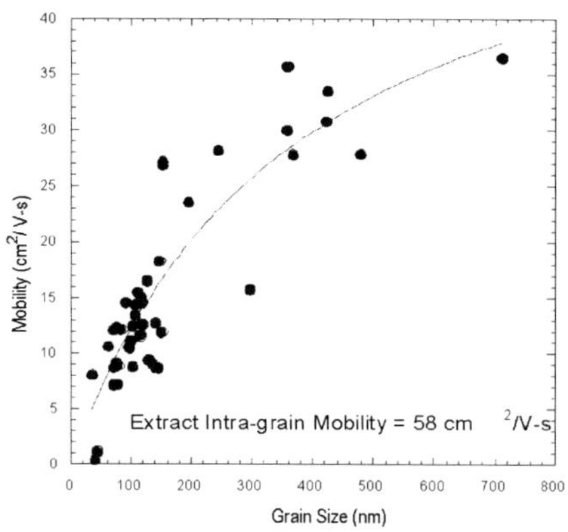

Figure.6 TFT mobility versus poly-Si grain size for poly-Si films crystallized by SPC, in the range of 600-800°C.

within this layer. Such concepts of stacked layers have been originally developed for furnace-annealing applications [24], [25] and are now successfully exploited in RTA crystallization, as well.

In figure 6 we have plotted the electron mobility as a function of the grain size of poly-Si films prepared by RTA (*this work*) or furnace anneal (*previous works*). As a general trend, it can be seen that the grain size correlates well with the mobility. However, the improvement of mobility with grain size tends to saturate beyond a certain point. This trend applies rather globally to poly-Si films prepared by SPC method (being furnace anneal or RTA). This trend can be explained by a mobility model that takes into account the effect of grain boundaries, as well as, that of defects within the grains [26]. For small grain size, the mobility is limited by the grain boundary energy barrier, whereas at large grain size, the intra-grain mobility dominates. Eq. 1 expresses the measured, total mobility as the sum of these two terms, ($\mu_{G\text{-}B}$→grain boundary mobility, $\mu_{I\text{-}G}$→ inter-grain mobility),

$$\frac{1}{\mu_{Total}} = \frac{1}{\mu_{G-B}} + \frac{1}{\mu_{I-G}} \tag{1}$$

The grain boundary related mobility term can be modeled as a function of grain size, as shown in eq. (2). For this derivation it is assumed that the grains are partially depleted [27]. The average grain size is represented by L_g, T is the absolute temperature, and v_c is defined as $(kT/2\pi m_e)^{0.5}$. E_B is the barrier height caused by trapped carriers at the grain boundaries. An expression for E_B is given in eq. (3), where N_T (cm^{-2}) represents the surface density of defects at a grain boundary, t_{ch} is thickness of the depleted poly-Si channel and V_{th} is the threshold voltage of the device. For a given crystallization method the grain boundary mobility will be approximately proportional to the grain size as implied by eq. (2), with a proportionality coefficient "K".

$$\mu_{G-B} = \left(\frac{qv_c}{kT} \cdot e^{\frac{-E_B}{kT}} \right) \cdot L_G = K \cdot L_G \tag{2}$$

$$E_B = \frac{q^3 N_T^2 t_{ch}}{8\varepsilon_s c_{ox} (V_{Gs} - V_{th})} \tag{3}$$

Substituting the equivalent relationship for $\mu_{G\text{-}B}$ ($=K \cdot L_G$) from eq. (2), eq. (1) becomes,

$$\frac{1}{\mu_{Total}} = \frac{1}{K \cdot L_G} + \frac{1}{\mu_{I-G}} \tag{4}$$

The experimental data in figure 6 were fitted with eq. (4) to yield best-fit values for the parameters $\mu_{I\text{-}G}$ and K. From this fit, the intra-grain mobility was extracted to be 61cm²/Vs and the average level of intra-grain defects in such poly-Si films was computed as $3 \times 10^{19} cm^{-3}$. The extracted intra-grain mobility places an upper bound as to the level of mobility performance that can be achieved in SPC poly-Si films crystallized in the temperature range of 600-800°C. Figure 7 shows a similar plot to figure 6 but with additional data from laser crystallized poly-Si films. The general trend is the same in all cases: the mobility initially increases with the grain size and eventually saturates due to intra-grain mobility limitations. The main difference between SPC and standard laser crystallization processes is the level of the intra-grain mobility. The model described by eq. (1)-(4) applies in both cases and suggests an intra-grain mobility, for laser annealed poly-Si films, in the range of 150-250cm²/Vs.

It is, therefore, concluded that the performance of SPC poly-Si films can never approach the high levels of excimer-annealed poly-Si films. However, moderate mobility values in the range of 20-40cm²/Vs with high uniformity can be useful in certain applications, such as pixel TFTs, especially in the field of Organic-Light-Emitting-Diodes. OLEDs are current-driven devices, thus, require high and uniform current at the pixel level. Hence, RTA technology provides an economic alternative to meet such requirements.

In the area of LCDs, SPC crystallization technology can also find suitable implementations.

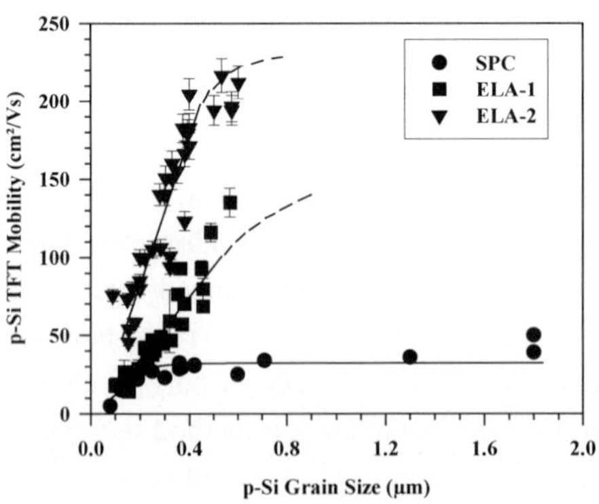

Figure.7 TFT mobility versus poly-Si grain size trend for SPC and laser annealed poly-Si films [28].

One such implementation relates to the combination of SPC and ELA technologies for the formation of high quality poly-Si films. When SPC is used with films doped with appropriate metal catalyst, the resulting poly-Si microstructure has improved characteristics, such as high crystal quality. Furthermore, the phase transformation process typically occurs at lower temperatures and at much faster rates than by conventional SPC method. If in addition, ELA process is combined with such improved microstructure, high performance material can be obtained as has been recently demonstrated [29].

3.Conclusions

Figure 8 summarizes the typical performance data for the key crystallization techniques discussed in this paper. Historically, solid-phase-crystallization has been the first method to produce poly-Si films for TFT applications. Since that time, around the early 80's, many other techniques have emerged, aiming to improve the film quality and TFT characteristics to enable new, more demanding display applications.

Excimer laser annealing has significantly improved and matured as a crystallization technique over the past 20 years. It is the only technique that has the potential of producing poly-Si TFTs with performance levels rivaling those of single-crystal Si. This high level of performance, however, has a high price tag and it is not necessary for every display application. RTA technology by itself offers very uniform, albeit modest performance levels which, however, are sufficient for certain applications (such as AMOLEDs). The development of mass production RTP equipment, compatible with glass substrates, has paved the way for the implementation of this technology in crystallization applications over the whole spectrum of display products. Either on its own, or in combination with other crystallization methods, RTA is expected to play an important role in the new poly-Si display era.

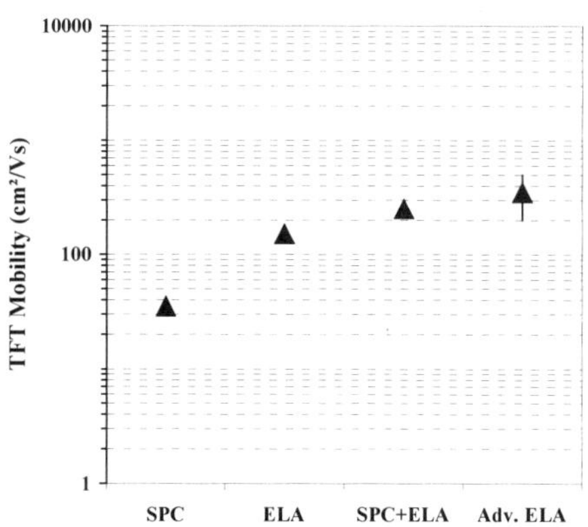

Figure.8 TFT mobility range versus poly-Si crystallization technique.

Acknowledgements.

The work at Lehigh University was supported by a grant from the National Science Foundation, NSF grant ECS-0088520.

References

[1] R. F. Cava et al., *Nature*, **377** (1995) 215.

[2] J.S. Im and H.J. Kim, *Appl. Phys. Lett.*, **63** (1993) 1969.

[3] A.T. Voutsas, C. Prat and D. Zahorski, *Proceedings of the 20th IRDC* (2000) 451.

[4] M. Fiebig, U. Stamm, P. Oesterlin, N. Kobayashi and B. Fechner, *Proceedings of SPIE*, **4295** (2001) 38.

[5] A. Demin et al., *Mat. Res. Soc. Symp. Proc.*, **695E** (2001)

[6] M. Matsumura and C.-H. Oh, *Thin Solid Films*, **337** (1999) 123.

[7] J.R. Koehler, R. Dassow, R.B. Bergmann, J. Krinke, H.P. Strunk and J.H. Werner, *Thin Solid Films*, **337** (1999) 129.

[8] J.S. Im and R.S. Sposili, *MRS Bulletin*, **21**(3) (1996) 39.

[9] S.W. Lee, Y.C. Jeon and S.K. Joo, *Appl. Phys. Lett.*, **66** (1995) 1671.

[10] S.Y. Yoon, S.K. Kim, J.Y. Oh, Y.J. Choi, W.S. Shon, C.O.Kim and J. Jang, Jpn. J. Appl. Phys., **37** (1998) 7193.

[11] K.C. Park, I.H. Song, S.H. Jung, J.W. Park and M.K. Han, *Proceedings of AMLCD 2000* (2000) 147.

[12] J.S Im, M.A. Crowder, R.S. Sposili, J.P. Leonard, H.J. Kim, J.H. Yoon, V.V. Gupta, H. Jin Song and H.S. Cho, *Phys. Stat. Sol. (A)*, **166** (1998) 603.

[13] K. Makihira and T. Asano, *Proceedings of AMLCD 2000* (2000) 33.

[14] H.J. Kim and J.S. Im, Appl. Phys. Lett., **68** (1996) 1513.

[15] P. van der Wilt, R. Ishihara and J. Bertens, *Mat. Res. Soc. Symp. Proc.*, **621** (2000) Q7.4.1.

[16] R.S. Sposili and J.S. Im, *Appl. Phys. Lett.*, **69** (1996) 2864.

[17] S.D. Brotherton, M.A. Crowder, A.B. Limanov, B. Turk and J.S. Im, *Proceedings of the 21st IRDC* (2001) 387.

[18] Y.H. Jung, J.M. Yoon, M.S. Yang, W.K. Park, H.S. Soh, H.S. Cho, A.B. Limanov and J.S. Im, *Mat. Res. Soc. Symp. Proc.*, **621** (2000) Q8.3.1.

[19] M. K. Hatalis and D.W. Greve, *J. Appl. Phys.,* **63** (1988) 2260.

[20] A. T. Voutsas and M. K. Hatalis, *J. Electrochem. Soc.*, **140** (1993) 871.

[21] A.T. Voutsas and M.K. Hatalis, K.R. Olasupo, A.K. Nanda and D. Alugbin, *Mat. Res. Soc. Symp. Proc.*, **345** (1994) 93.

[22] H. Hovagimian et al., *SID Proceedings Digest*, XXVIII (1997) M52.

[23] M. K. Hatalis et al., *Electrochem. Soc. Proc.*, **98-22** (1998) 92.

[24] A.T. Voutsas and M.K. Hatalis, *Appl. Phys. Lett.*, **63** (1993) 1546.

[25] A.T. Voutsas and M.K. Hatalis, *J. Appl. Phys.*, **76** (1994) 777.

[26] M. Stewart, *Ph.D. Dissertation*, Lehigh University (1999).

[27] J. Levinson, F.R. Shepherd, P.J. Scanlon, W.D. Westwood, G. Este and M. Rider, *J. Appl. Phys.*,

53 (1982) 1193.
[28] A.T. Voutsas, *to be published.*
[29] H. Sakamoto et al., *SID Proceedings Digest*, **XXXI** (2000) 1190.

Rapid Thermal Processing for Future Semiconductor Devices
H. Fukuda (Editor)

High-Performance Poly-Si TFT and its Application to LCD

Hiroki Hamada, Hisashi Abe and Yoshio Miyai

Microelectronics Research Center, SANYO Electric Co., Ltd., 180, Ohmori, Anpachi-cho, Anpachi-gun, Gifu, 503-0195, Japan
Phone: +81-584-64-4977 Fax: +81-584-64-5960 E-mail: hamada@ul.rd.sanyo.co.jp

1.Introduction

Poly-crystalline Si (poly-Si) Thin-Film Transistor (TFT) Liquid-Crystal Displays (LCDs) are very attractive for applications to apparatuses such as projectors, digital still cameras and personal computers [1-3]. As the electron mobility of poly-Si TFTs is 100 times higher than that of amorphous-Si (a-Si) TFTs, high-performance driver circuits can be monolithically integrated onto a panel [4]. Recently, to realize higher-performance poly-Si TFTs, several crystallization technologies, such as metal induced solid phase crystallization, metal induced lateral crystallization, metal imprint technology, excimer laser annealing (ELA) and stable scanning CW laser lateral crystallization have been developed [5-14]. However, no reports have yet been made on a high-temperature process that combines ELA and solid phase crystallization (SPC) and an activating technology for dopant using rapid thermal annealing (RTA).

In this paper, we report on a high-performance poly-Si TFT obtained by using a novel high-temperature (HT) process combined with SPC and ELA, and on the characteristics of a poly-Si TFT LCD light valve with fully integrated circuits fabricated using the resulting TFT. We also report on a novel activation process under low-temperature (LT) conditions combined with RTA and an ion doping (I/D) method, which is required to mass-produce the LT poly-Si TFT LCDs.

2.Experiments

2-1 TFT process flows

Fig. 1 shows the process flows of top-gate TFTs. The left and right sides shows the HT process and the LT process, respectively. In the HT process, amorphous Si (a-Si) films were formed on fused quartz substrates. The a-Si films were crystallized by ELA and/or SPC methods. Then, poly-Si films were annealed by various methods, as shown in Table I. After patterning the poly-Si films, gate insulating films (SiO_2) and a gate electrode (n+ poly-Si films) were formed on the poly-Si films. Ion Implantation (I/I) was performed after patterning the gate electrode. To activate the dopants, a furnace annealing method was applied. Then, an insulating layer was deposited on the poly-Si film using the LPCVD process and contact holes

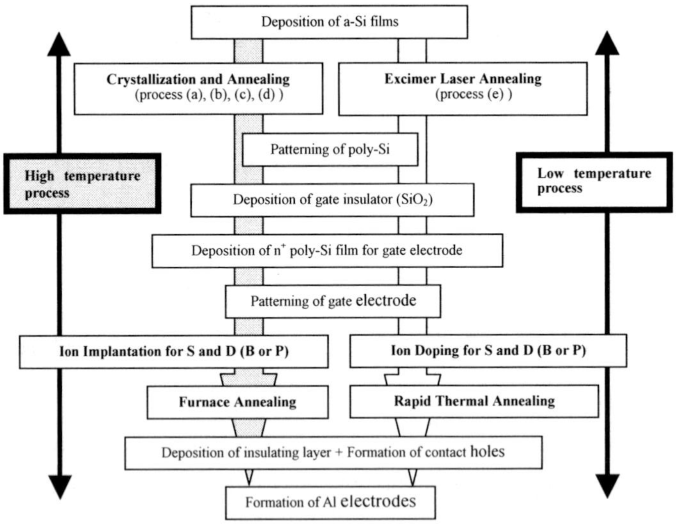

Symbols (a) – (e) shows samples which are prepared by several annealing process in Table I.

Fig.1 Process flows for top- gate TFTs

were formed by the dry etching method. Finally, Al electrodes were deposited by the sputtering method [15].

Compared with the HT process, the LT process has two mail differences. One is the crystallization of the a-Si films by the ELA method. Another is the activation of the dopant due to form the source and drain. The conditions for these processes are described in the following section.

The surface roughness of the poly-Si films was measured using atomic force microscopy (AFM) method. The crystallinity of poly-Si films was investigated using electron spin resonance (ESR), transmission electron microscopy (TEM) and the transmission electron diffraction (TED) methods.

2-2 High temperature process

Poly-Si films were prepared by using various annealing processes, as shown in Table I . Samples (a), (b), (c1), (c), (d) and (e) were prepared by combining SPC and the thinning process; combining ELA and the PHT process; combining SPC, ELA and the thinning process; combining SPC, ELA, the thinning process and the PHT process; SPC; and ELA, respectively. TFTs were fabricated using the processes shown in Fig. 1.The a-Si films were formed on fused quartz substrates using a low-pressure chemical vapor deposition (LPCVD) method at a deposition temperature of 450°C for all samples. The source material was Si_2H_6 gas. SPC was done to crystallize the a-Si films at 600°C in ambient of N_2 for 20 hours. The thinning process was introduced to reduce the defects in the inter-grain. The process was

Table I Annealing methods for forming poly-Si

Samples	Annealing methods
(a)	SPC + thinning
(b)	ELA + PHT
(c1)	SPC + thinning + ELA
(c)	SPC + thinning + ELA + PHT
(d)	SPC
(e)	ELA

performed in ambient of O_2 at 1000°C for 0.5 hours. The ELA process was used to reduce the potential barrier height enhanced by the crystallization defects at the grain boundary and/or inter-grain. The process was performed using KrF excimer laser with the substrate at room temperature in a vacuum chamber. The beam size was 0.5×50 mm and the overlap of the beam was 90%. The energy density of the ELA ranged from 200 to 400 mJ/cm^2. A post-heat treatment process (PHT) was carried out in ambient of N_2 at 1000°C for 2 hours. The PHT process is applied to reduce the surface roughness and crystalline defects of poly-Si films that were enhanced by the ELA process [15].

Fig. 2 shows the TEM images, and the TED patterns of poly-Si films for samples (a), (b), (c) and (d), respectively, are shown in Table I. Each TED photograph shows typical patterns that were observed at the inter-grain of the samples. The observed location is shown by the arrow in the TEM image. The ELA energy density was approximately 350 mJ/cm^2. The TEM images show that the grain size of samples (a), (c) and (d) were almost the same and large. On the other hand, sample (b) is characterized by a small grain size (<0.07 μm) and is very different from samples (a), (c) and (d) in the TEM image. From the TED patterns of (a) and (c), we found that the number of crystalline defects with the inter-grain was drastically decreased by introducing the ELA and PHT processes.

Table II summarizes the characterization of poly-Si films (a) to (e) in Table I. The energy density of the ELA process was approximately 350mJ/cm^2. From the results of spin densities

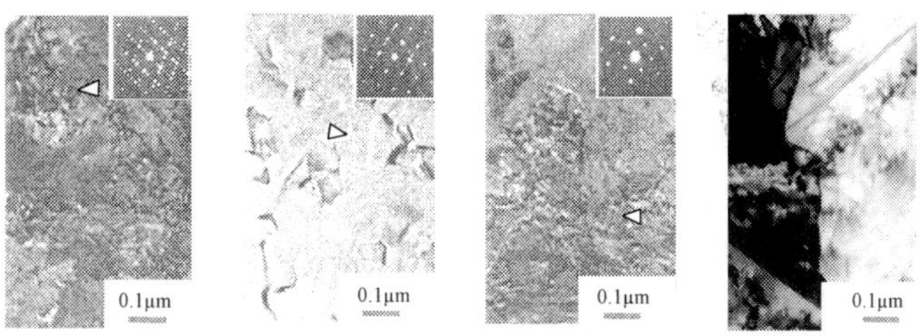

(a) SPC+thinning (b) ELA+PHT (c) SPC + thinning + ELA + PHT (d) SPC

Fig.2 EM images and TED patterns of poly-Si films

Table II Characteristics for poly-Si films

Samples	Spin densities ($\times 10^{17}cm^{-3}$)	Grain size (μm)	Rrms (nm)
(a)	<1.0	2.0	1.0
(b)	5.0	0.07	1.0
(c1)	32	2.0	5.0
(c)	<1.0	2.0	1.0
(d)	34	2.0	1.0
(e)	30	0.07	6.0

for sample (a) and (d), the thinning process was very useful for obtaining poly-Si films with low crystalline defects. In this experiment, we tried to obtain higher-quality poly-Si films. From the characteristics of sample (c), we found that high-quality poly-Si films were successfully fabricated by a novel process that combines SPC, thinning, ELA and PHT. The characteristics of sample (c) were the same as those of sample (a) in Table I . However, the quality of sample (c) was better than that of sample (a), from the results of TED patterns in Fig.2 (a), (c). We found that the PHT process was very useful for reducing surface roughness and the crystalline defect in poly-Si films in comparison with sample (b) and (e) or (c1) and (c).

Fig. 3 shows the I_{DS}-V_{GS} characteristics for the HT-processed poly-Si TFTs of samples (A), (B) and (C), which were fabricated from poly-Si films (a), (b) and (c) in Table I . μ_{FE}, V_{TH} and S show the field effect mobility, threshold voltage and sub-threshold swing for the TFT, respectively. The performance of TFT (C) was much better than that of samples (A) and (B). Taking into account the fact that the TFT processing procedure of this experiment was basically the same, the crystalline quality of poly-Si film (c) is superior to those of poly-Si films (a) and (b). From these results, we found that a method combining the ELA and PHT process was very useful in obtaining high-performance poly-Si TFTs [15].

2-3 Low temperature process

In the LT process of poly-Si TFTs, the process for activating the dopants has been

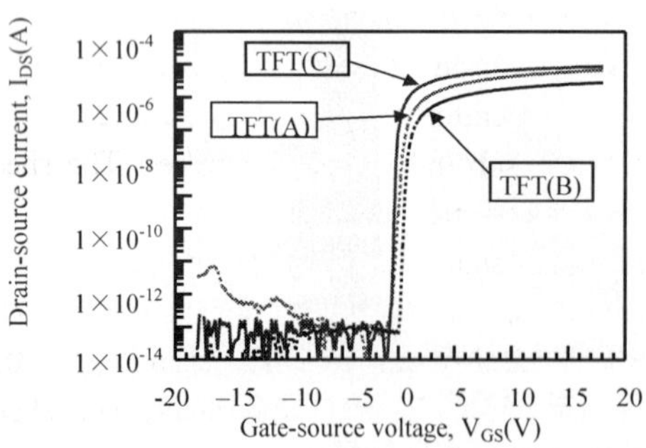

TFTs	μ_{FE} ($cm^2/V\cdot s$)	V_{TH} (V)	S (V/dec.)
(A)	150	0.92	0.30
(B)	72	0.87	0.12
(C)	325	0.15	0.08

Fig.3 I_{DS}-V_{GS} characteristics of poly-Si TFTs

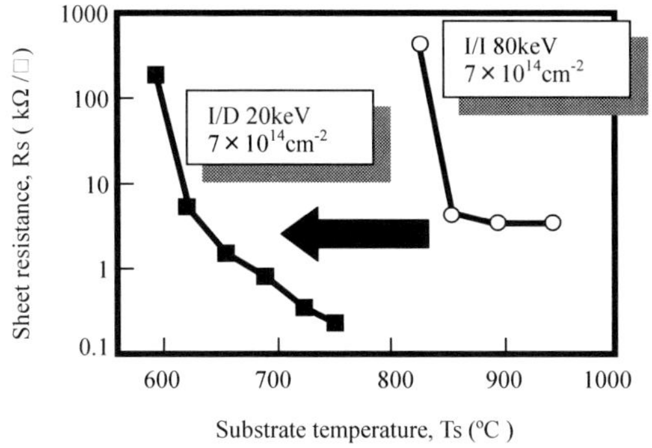

Fig.4 Sheet resistance for P-doped poly-Si films prepared by the I/D and I/I methods versus substrate temperature during activation process by RTA

important for obtaining high-performance poly-Si TFTs. The dopants have been mainly activated using the ELA method. However, this results in lower throughput and lower uniformity than the RTA method. On the other hand, RTA method required a higher process temperature than ELA method, and could not be applied to activate dopants in poly-Si films prepared on glass substrates.

Recently, we have developed a novel activation technology combined with RTA and the ion doping method for LT-processed poly-Si TFT. Phosphorus (P)-doped poly-Si films with a lower sheet resistance, which are suitable for poly-Si TFT LCDs, could be obtained on glass substrates using this novel process. This process is very useful for obtaining poly-Si TFTs with uniform performance on a glass substrate with large area.

Amorphous-silicon (a-Si) films (thickness: 50 nm) were formed on Corning #1737 glass substrates with a SiO_2 buffer layer using an LPCVD method at a growth temperature of 400°C. The a-Si films were crystallized by ELA at an irradiation power of 300 mJ/cm^2. P-ion doping was done at an ion beam energy of 20 keV. P-ion implantation was also done at a beam energy of 80 keV through a 100 nm thick oxidization film. The amount of the dose in the poly-Si films was 7×10^{14} cm^{-2}.

The RTA apparatus was mainly constructed of two Xe arc lamps with focusing reflectors and pre-heating plates. The focus point of the reflectors brought onto the surface of the glass substrate to increase the annealing effect. The temperature of the glass substrates was observed by pyrometers. The transfer velocity of the substrates was 12 mm/sec. The sheet resistance for the Si films was examined by the four points probe measurement.

Fig. 4 shows the relationship between the sheet resistance of P-doped poly-Si films (R_S) and the substrate temperature (Ts) during RTA process. For phosphorus doping into poly-Si films, the ion implantation (I/I) or ion doping (I/D) method was used. The shrinking of the Corning #1737 glass substrate happened at about 850°C in this experiment. The sheet resistance of the P-doped poly-Si films decreased with an increase in the substrate temperature during RTA process. The sheet resistance of the ion doped poly-Si films was

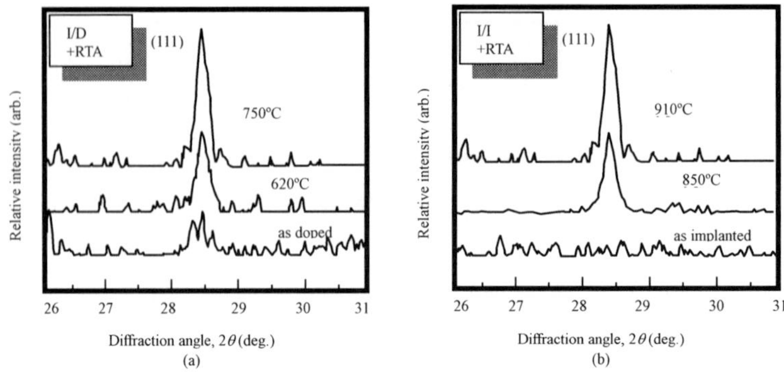

Fig.5 X-ray diffraction patterns for P-doped Si films as function of substrate temperature during activation process by RTA. X-ray diffraction patterns (a) and (b) are for P-doped Si films prepared by the I/D and I/I method, respectively.

more than 10 times lower than that of films by the I/I method, and obtained 180 Ω/□ by RTA method at 750°C. The substrate temperature that resulted in poly-Si films with a sheet resistance of 3 kΩ/□ was 620°C for the films doped by the I/D method and 850°C for the ion implanted films.

Fig. 5 shows the X-ray diffraction patterns for P-doped Si films prepared by the I/D (a) and I/I (b) method as a function of the substrate temperature during RTA process. With an increase in the substrate temperature, P-doped amorphous or microcrystalline Si films were converted into a (111) plane oriented polycrystalline structure. The crystalline temperature for the P-doped Si films prepared by the I/D method was about 230°C lower than that for the I/I method. We believe that the activation of the P-doped films and the crystallization were effectively achieved at low temperature by combining RTA and the I/D method.

Fig. 6 shows photographs of transmission electron microscopy (TEM) for the P-doped poly-Si films having a dose of 1×10^{15} cm^{-2} by the I/D and the I/I methods. These films were not subjected to RTA. The doped Si films prepared by the I/I method were related to the amorphous films from the polycrystalline films, as shown in Fig. 6 (b). On the other hand, the

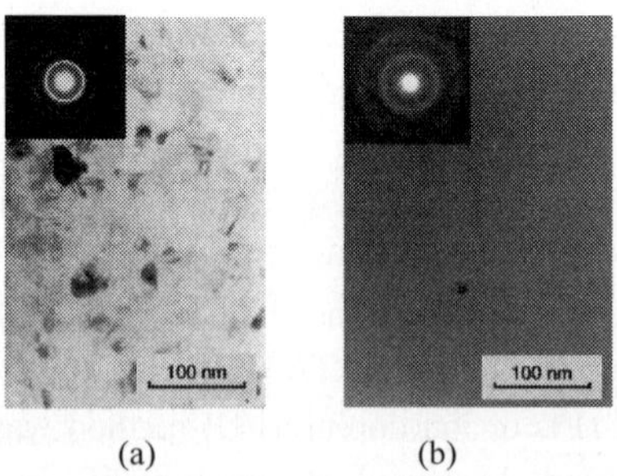

Fig.6 TEM photographs for P-doped Si films having a dose of $1\times10^{15}cm^{-2}$. Photograph (a) and (b) are P-doped Si films prepared by the I/D and I/I method, respectively

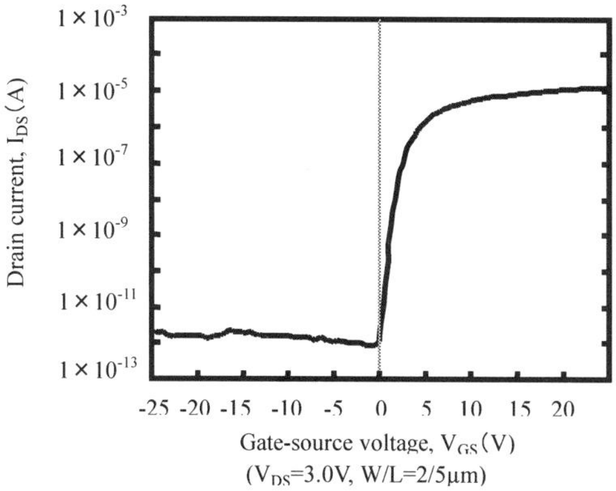

μ_{FE} (cm^2/V·s)	V_{TH} (V)	S (V/dec.)
70	2.0	0.3

Fig.7 I_{DS}-V_{GS} characteristics for n-ch poly-Si TFT with LDD structure

doped films prepared by the I/D method were observed, in which the polycrystalline structure remained in the films, as shown in Fig. 6 (a). From these results, we think that the activation of the P-dopant and the crystallization at lower temperature were accelerated by the polycrystalline nucleus that remained in the Si films. We thus found that P-doped poly-Si films could be obtained with low resistance without shrinking for the glass substrates.

Fig. 7 shows the I_{DS}-V_{GS} characteristics for low-temperature processed poly-Si TFT with a lightly doped drain (LDD) structure of TFT (E), which was fabricated from the poly-Si film (e) in Table I. The length of the LDD was 1.0 μm. The activation of the source and drain region for this TFT was carried out with the novel RTA described in this section. We obtained good transfer characteristics in the TFT using ELA and the novel RTA. This performance was sufficient to create TFTs for the pixels of an LCD [16].

2-4 Application of poly-Si TFTs

In this section we introduce the application of high-temperature poly-Si TFTs. Fig. 8

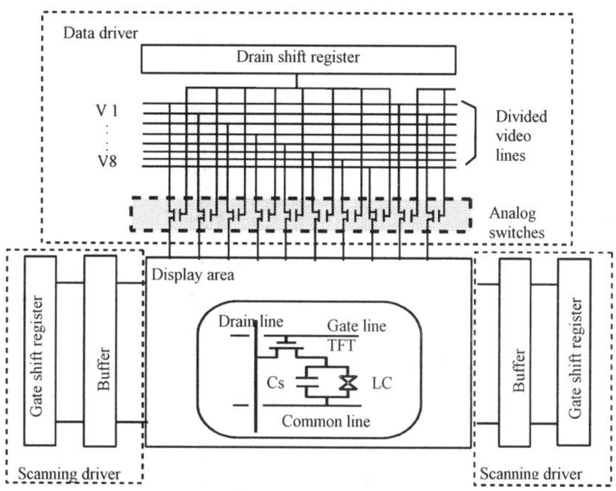

Fig.8 Circuit diagram of an LCD light valve for HDTV

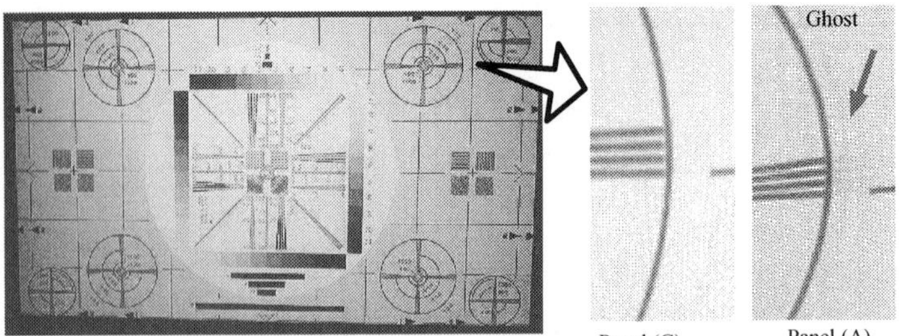

ITE chart

Panel (C)
(SPC + thinning + ELA)

Panel (A)
(SPC + thinning)

Fig.9 Projected image of LCD light-Valve

shows the circuit diagram of an LCD light valve for high-definition television (HDTV) fabricated using HT-processed poly-Si TFT [18]. The data and scanning drivers are monolithically integrated onto a fused quarts substrate. The data lines were operated by driver circuits that consisted of shift registers and analog switches. To sample an HDTV video signal, the 1440 video lines were divided into eight phases. This is why the bandwidth of the TFTs is about 10MHz. Each line was connected to the data lines through analog switch TFTs. The TFTs for analog switches conducted the sampling and holding of the video signal. The shift registers in the data and scanning drivers consisted of C-MOS inverter circuits. The structure and panel process for this light valve have been reported elsewhere [17].

Fig. 9 shows screen-displayed images of projectors using two different panels. The panel (A) and (C) were fabricated from poly-Si films (a) and (c) in Table I , respectively. The ghost in panel (C) was smaller than that in panel (A). In general, ghosts depend on the steepness of the video signal, the delay time in the data driver, and the switching speed of analog switches. Therefore, we believe that the delay time in the data driver and the switching speed of analog switches are drastically improved using high performance TFTs [18].

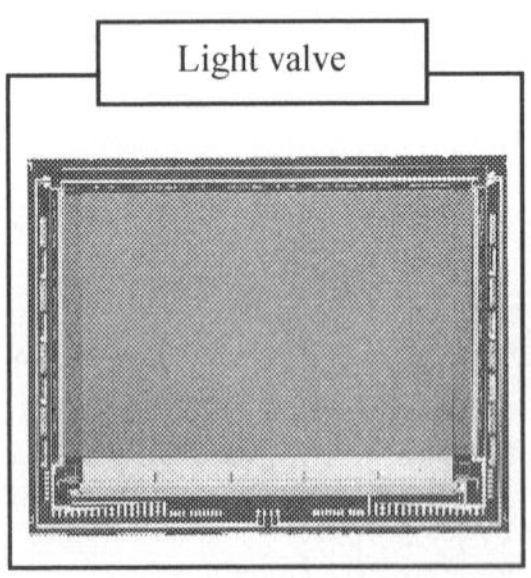

Display size	6.35cm (2.5”) diagonal
Number of pixels	1440 (H)×1024 (V) (1.5M)
Pixel pitch	38.5 (H)×30.5 (V) μm
Peripheral drivers	Both scanning and data drivers
Number of video input	8
Scan mode	Two lines simultaneously and
Aperture ratio	cross inversion driving (CID)
Contrast ratio	35%
Horizontal resolution	>900:1 (on panel)
Mirror inversion	810 TV lines By changing scan direction of

Fig.10 High-definition poly-Si TFT LCD light valve module

Fig. 10 shows a high-definition (1.5 mega-pixels) poly-Si TFT LCD light valve module. The display size and the number of pixels were 2.5" (diagonal) and 1440 (H) × 1024 (V) (1.5 mega-pixels), respectively. The pixel pitch is 38.5 (H) × 30.5 (V) μm. The shape of the display area was 16:9, which is suitable for the display of HDTV. We also achieved a high contrast ratio of 900:1 and resolution of 810 TV lines using high-performance poly-Si TFT.

3.Conclusion

We have obtained high-quality poly-Si films by a novel high-temperature processed crystallization process, which combines the SPC, ELA, and PHT processes. High-performance TFTs were fabricated using this method. The field effect mobility, threshold voltage and sub-threshold swing for the poly-Si TFTs were 325 $cm^2/V\cdot s$, 0.15 V and 0.08 V/dec., respectively. We applied these high-performance TFTs to an HDTV projector, and showed that the reduction of ghost can be realized.

We have also successfully developed a low-temperature activating technology, which was combined with the RTA and I/D method. P-doped poly-Si films with a sheet resistance of 3 k Ω/□ were obtained at a substrate temperature of 620°C without shrinking of the glass substrates. The process temperature of the novel technology was 220°C lower than that for the conventional process. We could obtain a low-temperature processed poly-Si TFT with LDD structure using the novel activating technology. The field effect mobility of the TFTs is as high as 70 $cm^2/V\cdot s$. This value is enough to fabricate simple driver circuits for LCD.

We hope that LCDs known as "System on a Panel" will be developed using technologies based on our proposed process in the near future.

References

[1] Y. Hayashi, M. Iida, T. Iwanaga, H. Kaneko, S. Nishihara, H, Kitagawa and S. Koike, *Digest of Technical Papers of AM-LCD96*, (1996) 157.

[2] K. Uchino, T. Kashima, F. Abe, M. Hirano, M. Satoh and T. Maekawa, *SID'98 Digest*, (1998) 1067.

[3] N. Ibaraki, *SID'99 Digest*,(1999) 172.

[4] H. Kimura, T. Maeda, T. Tunashima, N. Moriyama, T. Morita and H. Sato, *SID'01 Digest*, (2001) 268.

[5] N. Matsuo, Y.Aya, T. Kanamori, T. Nouda, H. Hamada and T. Miyoshi, *Jpn. J. Appl. Phis., vol.39, Pt.1, No.2A*, (2000) 351.

[6] N. Kawamoto, H. Abe, N. Matsuo, R. Taguchi, T. Nouda and H. Hamada, *Digest of Technical Papers of AM-LCD01*, (2001) 127.

[7] L. Mariucci and G. Fortunato, *Digest of Technical Papers of AM-LCD01*, (2001) 135.

[8] T. K. Kim, B. I. Lee, C. H. Park, H. Ho. Sung, G. B. Kim, W. S. Shin, J. S. Lee and S. K. Joo, *Digest of Technical Papers of AM-LCD97*, (1997) 187.

[9] K. C. Saraswat and V. Subramanian, *EDF'98 proceedings, (1998) 2.*

[10] J. Koyama, S. Nakajima, H. Ohnuma, Y. Shionoiri and S. Yamazaki, *SID'98 Digest*, (1998) 883.

[11] K. Makihira, M. Yoshii and T. Asano, *Digest of Technical Papers of AM-LCD01*, (2001) 235.

[12] M .A. Crowder, P. G. Carey, P. M. Smith, S. Sposili, H. S. Cho and J. S. Im, *IEEE Electron Device Lett., vol.19,No. 8*, (1998) 306.

[13] C. H. OH and M. Matsumura, *Digest of Technical Papers of AM-LCD99*, (1999) 255.

[14] A. Hara, F. Takeuchi, M. Takei, K. Yoshino, K. Suga and N. Sasaki, *Digest of Technical Papers of AM-LCD01*, (2001) 227.

[15] H. Hamada, Y. Aya, H. Abe, T. Nouda, K. Hirano and Y. Miyai, *Proceedings of Thin Film Transistor Technologies IV*, (1999) 11.

[16] K. Hirano, N. Sotani, I. Hasegawa, T. Nouda, H. Abe and H. Hamada, *SPIE Vol. 3014, Proceedings of Active Matrix Liquid Crystal Displays Technology and Applications*, (1997) 119.

[17] Y. Aya, T. Nouda, H. Abe, Y. Miyai, M. Iseki and H. Hamada, *Digest of Technical Papers of AM-LCD99*, (1999) 25.

[18] Y. Aya, D. Ide, H. Abe, Y. Miyai, M. Iseki and H. Hamada, *IDW'99 Digest*, (1999) 1093.

Rapid Thermal Processing for Future Semiconductor Devices
H. Fukuda (Editor)

Rapid Low Temperature Photo Oxidation Processing for Advanced Poly-Si TFTs

Yukihiko Nakata [1], Takashi Itoga[2], Tetsuya Okamoto[2], Toshimasa Hamada[2] and Yutaka Ishii[3]

[1, 2]SHARP Corporation, System LCD Group Process Development Laboratories
2613-1, Ichinomoto-cho, Tenri, Nara, 632-8567, Japan
Phone:+81-743-65-1321 Fax:+ 81-743-65-3425
3SHARP Corporation, Display Development Group Display Laboratories.
[1]Present Address: Advanced LCD Development Center Co.,Ltd. Dep. 1
292, Yoshida-cho, Totsuka, Yokohama, Kanagawa, 244-0817, Japan
Phone:+81-45-866-6419 Fax:+81-866-6405 E-mail:nakata-ALT@center.email.ne.jp

1.Introduction

The mobile display market has been changing with the evolution of the information technology. The display is trying to be more compact and more by integrating new functions. The Low Temperature Poly Silicon (LPS) TFT is one of the most expected technologies to realize the integrated circuits on the display panels. The gate insulator formation is one of the key technologies for low temperature Poly-Si TFTs. The gate insulator for the Poly-Si TFT needs to form less than 600°C and to cover on the large glass substrate compared with the gate insulator for LSI. The SiO_2 gate insulator, which is deposited by PECVD method using TEOS gas source, is most popular. However, SiO_2 films using TEOS gas needs annealing at about 600°C, 2h to improve the film density after deposition. The interface characteristics of SiO2/Si are not enough to get the small deviation of the threshold voltage.

The oxidation process can form the excellent SiO_2/Si interface, because the interface is formed inside the original Si. However, the thermal oxidation cannot apply to the low temperature Poly-Si TFTs because the thermal oxidation rate is too slow under 600°C. Therefore, the low temperature oxidation is important for Poly-Si TFTs. Fuyuki et al. proceeded the oxidation using the oxygen radical by remote plasma [1]. Ohmi et al. proposed Kr/O_2 plasma oxidization [2]. The characteristics of the Kr/O_2 plasma oxide do not depend on the crystal orientation. Therefore, this method is fit to Poly-Si TFT. Also, photo oxidation was researched, but the oxidation rate is low [3]. These oxidation methods are researched for the application to LSI and total film thickness are less than 5 nm. However, these thin films cannot apply to Poly-Si TFTs because Poly-Si surface has large roughness and the area of glass is more than 9 times larger than Si single crystal wafers. Therefore, the objects of this R&D are to get the excellent interface characteristics of SiO_2/Si and gate insulator at low temperature and eliminate the annealing process after the gate insulator formation.

2.Combination of Photo Oxidation and PECVD

We propose the combination process of the low temperature photo oxidation and PECVD for the gate insulator for Poly-Si TFTs to obtain the excellent interface and the sufficient film thickness to cover the large glass substrate at low temperature. The photo oxidation has the big advantage of no ion bombardment compared with the plasma oxidation. Si single crystal wafers were used to evaluate the SiO_2/Si characteristics. As shown in Fig. 1, Si wafer were set and the chamber was evacuated. Oxygen gas was introduced and was kept the constant presser. The 172 nm lights from Xe excimer lamp was irradiated to the oxygen gas to make the active oxygen radicals O(1D). They were generated directly and efficiently without the process of the ozone. The oxygen radicals are very active and make the SiO_2 efficiently.

There is the optimum oxygen gas pressure to obtain the maximum photo oxidation rate. The generation of the oxygen radical is reduced in the case of low gas pressure. In the case of high gas pressure, the 172 nm light is absorbed at the quite near the lamp and far from the substrate and oxidation rate is reduced.

After 3nm photo oxidation, the second SiO_2 films were deposited by PECVD from SiH_4 + N_2O gas. The excellent SiO_2/Si interfaces with the interface trap density of 2-3E+10/cm^2/eV was obtained by the photo oxidation at 200-300°C.

This interface trap density is the same as it of the thermal oxidation at 950°C and about 1/4 of TEOS SiO_2 with 600°C, 2h anneal. However, in the case of PECVD second films from SiH_4 + N_2O, the flat band voltages (Vfb) were changed by the stress test. It may be caused by the remaining nitrogen in the film.

In order to improve the reliability without nitrogen impurity, the second SiO_2 films were deposited by PECVD from TEOS (Tetra Ethyl Ortho Silicate) + O_2 after 3 nm photo oxidation. The Vfb and Vfb shift after stress test are shown in Fig. 2. As the total film thickness was reduced, Vfb and Vfb shift were reduceed according to the ordinal rule. The stacked film of 3 nm photo oxide and 40 nm PECVD film from TEOS without anneal has the same Vfb and Vfb shift of 100 nm PECVD film from TEOS with 600°C anneal, in spite of the half film thickness. The film has the excellent interface trap density 3E+10/cm^2, as sown in Fig. 3. Also,

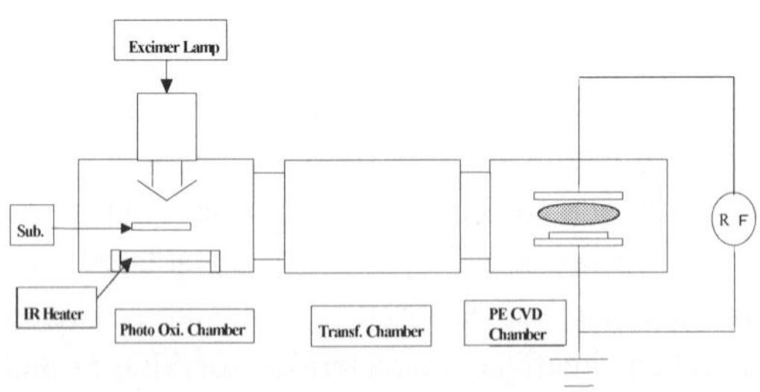

Fig.1 Equipment for Photo oxidation and PECVD.

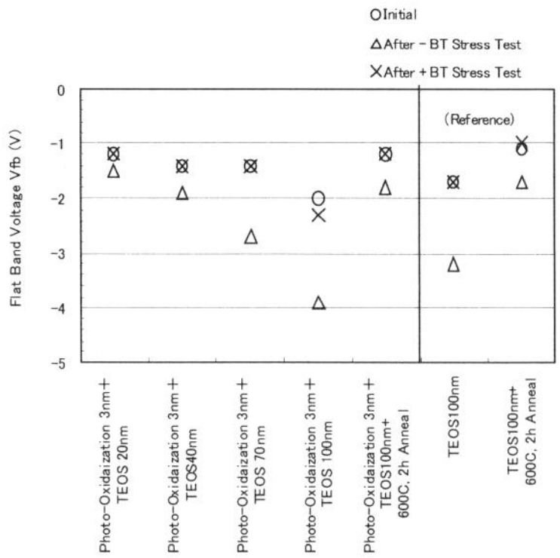

Fig.2 Flat Band Voltage of Photo oxidization + PECVD

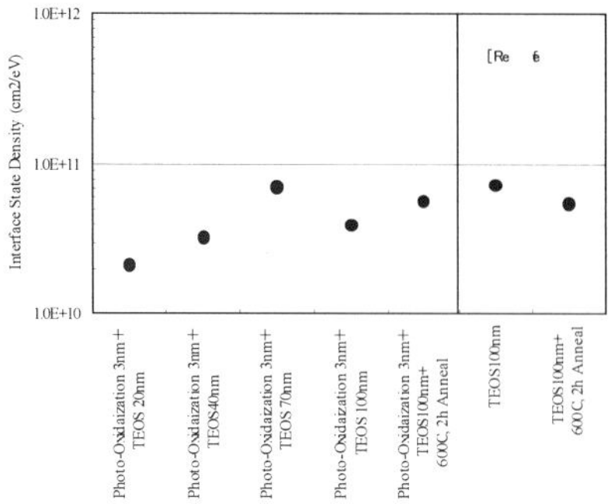

Fig.3 Interface State Density of Photo oxidization + PECVD (TEOS+O_2)

the film has the leak current less than 1E-10 A/cm^2 and the breakdown electric field more than 8MV/cm as shown in Fig. 4.

3.Conclusions

The SiO_2 stacked insulators, whose 2nd layers were deposited by PECVD after forming the good SiO_2/Si interface layer by photo oxidation at low temperature, were studied.

As the results, the following conclusions were obtained.

The excellent SiO_2/Si interfaces with the interface trap density 2-3E+10/cm^2/eV was obtained by the photo oxidation using Xe excimer lamp at 200-300C. This interface trap density is the same as it of thermal oxidation at 950°C and about 1/4 of current TEOS SiO_2 with 600°C, 2h anneal.

The staked film of 3 nm photo oxide and 40 nm PECVD film from TEOS without anneal has the same characteristics, such as leak current, breakdown electrical field and reliability, of 100 nm PECVD film from TEOS with 600°C anneal, in spite of the half film thickness.

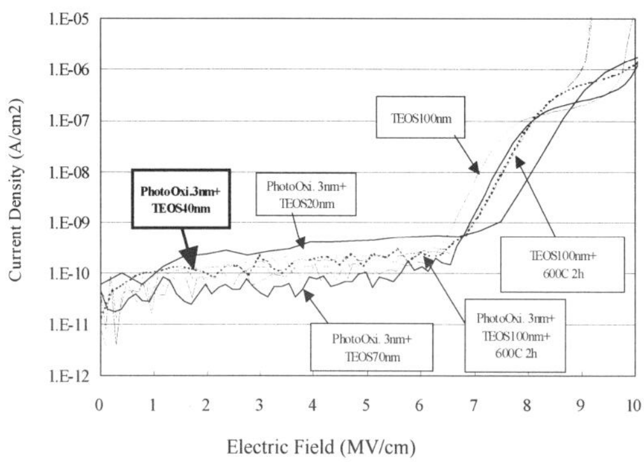

Fig.4 I-V Characteristics of Photo oxidization + PECVD

This photo oxidation and PECVD combination process make the low temperature high quality gate insulator for Poly-Si TFTs under 300°C. This technology leads the Poly-Si TFTs to not only high performance but also high mass production yields with small deviation of threshold voltages in order to realize the system display integrated the high functions.

Acknowledgements

The photo oxidization is proceeded as the part of the research that is entrusted to ASET by NEDO.

References

[1] M. Hirayama et al., IEDM Tech. Dig., (1999) 249.
[2] T.Fuyuki, T. Oka and H. Matsunami, J. J. A. P. 33(1994) 440.
[3] Jun-Ying Zhang and Ian W. Boyd, A. P. L. 71(20)1997 2984

Properties of Phosphorus-Doped Polycrystalline Silicon Films Formed by Catalytic Chemical Vapor Deposition and Successive Rapid Thermal Annealing

Rui Morimoto, Akira Izumi, Atsushi Masuda and Hideki Matsumura

School of Materials Science, Japan Advanced Institute of Science and Technology (JAIST), Tatsunokuchi, Ishikawa 923-1292, Japan
Phone: +81-761-51-1564, Fax: +81-761-51-1505, E-mail: rmorimot@jaist.ac.jp

1.Introduction

Phosphorus (P) doped polycrystalline silicon (poly-Si) thin films are used as gate or capacitor electrodes for ultralarge-scale integrated circuits (ULSIs) and source/drain electrodes for thin-film transistors (TFTs) for active-matrix liquid-crystal displays (AMLCDs). However, the deposition temperature for conventional low-pressure chemical vapor deposition (LPCVD) is around 600°C or higher. Further reduction of thermal budget is desirable for forming shallow junction in ULSIs and for avoiding substrate damages in AMLCDs.

It is known that amorphous silicon (a-Si) films prepared by catalytic chemical vapor deposition (Cat-CVD) contain a few amounts of hydrogen (H) [1]. This is an advantage to transform a-Si films to poly-Si films by annealing since it is expected to eliminate H bubbling causing rough surface. Therefore, poly-Si films with both low resistivity and perfect flatness are expected using Cat-CVD process.

In the present work, properties of P-doped poly-Si films prepared by Cat-CVD and successive rapid thermal annealing (RTA) were investigated.

2.Experiments

Figure 1 shows schematic diagram of the Cat-CVD apparatus. A tungsten (W) wire in 0.5 mmϕ and 1650 mm-length was used as a catalyzer. It was set at an area of 125×125 mm^2. The distance between the catalyzer and the samples was fixed at 40 mm. Film thickness was controlled by adjusting deposition time with a shutter located between the catalyzer and the samples. The catalyzer temperature was monitored based on the resistance of the W wire and also radiation power observed by an infrared pyrometer. The substrate-holder temperature monitored by a thermocouple was controlled by a heater inside the substrate holder. The process chamber was evacuated by a turbo molecular pump (TMP) below 8.0×10^{-6} Pa. The gas pressure was adjusted by the rotation rate of TMP during the process. A gas mixture of silane (SiH_4) and phosphine (PH_3) gases, diluted at 2% by helium (He), was used for the

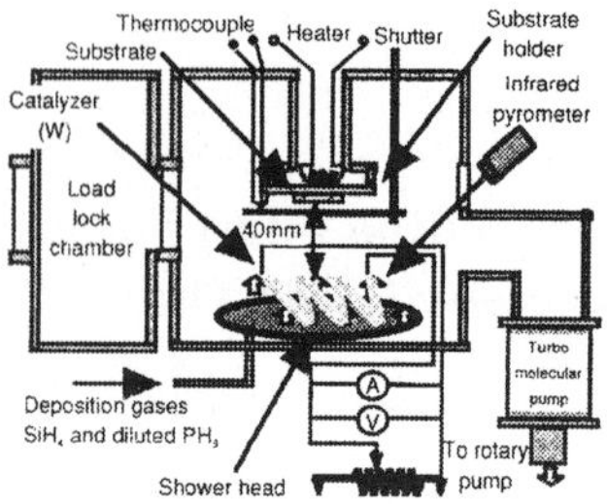

Fig.1 Schematic diagram of Cat-CVD apparatus.

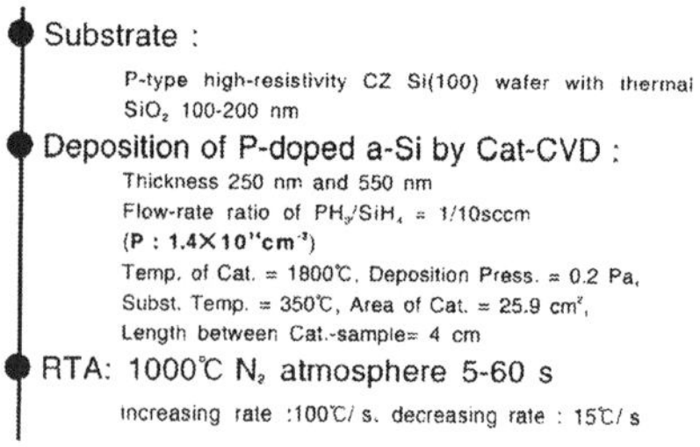

Fig.2 Process flow of the sample preparations.

deposition of P-doped a-Si films.

Figure 2 shows process flow of the sample preparations. Here, we chose the deposition condition of the P sheet concentration of 1.4×10^{14} cm^{-2}, which was realized at flow-rate ratio of PH_3 to SiH_4 of 0.1. In total-reflection X-ray fluorescence (T-XRF) measurements, the X-ray penetrates into the films at about a 10 nm-depth. The integral of P sheet concentration from the surface to a 10 nm-depth should be detected by T-XRF, and thus, the three-dimensional P concentration will be over 10^{20} cm^{-3}. a-Si films were annealed by RTA in N_2 atmosphere at 1000°C for crystallization. Additionally, furnace annealing (FA) was also carried out in N_2 atmosphere at 1000°C for 30 min to investigate for comparison.

The films were characterized by Raman spectroscopy for estimation of the crystalline fraction. Cross-sectional scanning electron microscopy (SEM) and atomic force microscopy (AFM) were used for the observation of the surface morphology. The grain size was evaluated by XRD. The resistivity was measured using coplanar electrodes for a-Si films and by the van der Pauw method for poly-Si films, because the resistivity of a-Si films was too high to measure by the van der Pauw method. Finally, metal-oxide-semiconductor field effect transistors (MOSFETs) were fabricated using these poly-Si films as gate electrodes to determine the feasibility of device application of the present poly-Si films. Gate-leakage current and capacitance-valtage (C-V) characteristics were also measured. In particular, C-V measurements were carried out at the frequency of 1 kHz and sweep rate of 0.1 V/s to estimate the level of the gate-depletion effect.

3.Results and Discussion

H contents in Cat-CVD a-Si films

To investigate the feasibility of Cat-CVD a-Si films as precursor films for solid phase crystallization (SPC), the H contents in the films were estimated by Fourier-transform infrared (FTIR) method. Figure 3 shows FTIR spectrum. We used 2.1×10^{19} cm^{-2} for Si-H wagging vibration at 640 cm^{-1} [2]. The H content in the films was estimated to be 1.1 at.% and it is much lower than that for PECVD a-Si films of over 10 at.%. Therefore, the films are expected to be suitable for SPC processes. Figure 4 shows optical microscopic photographs of surface

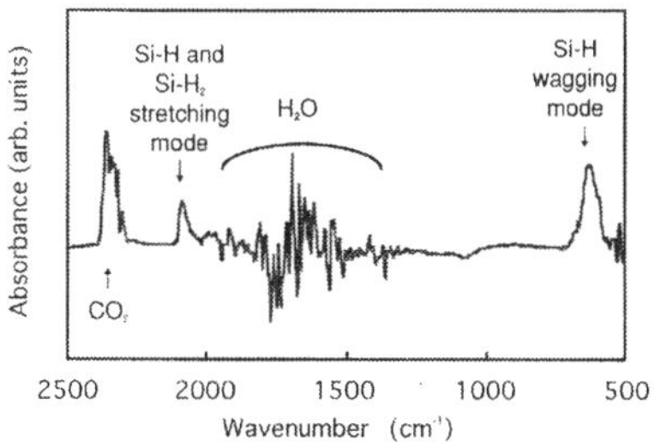

Fig.3 FTIR spectrum of Cat-CVD a-Si film.

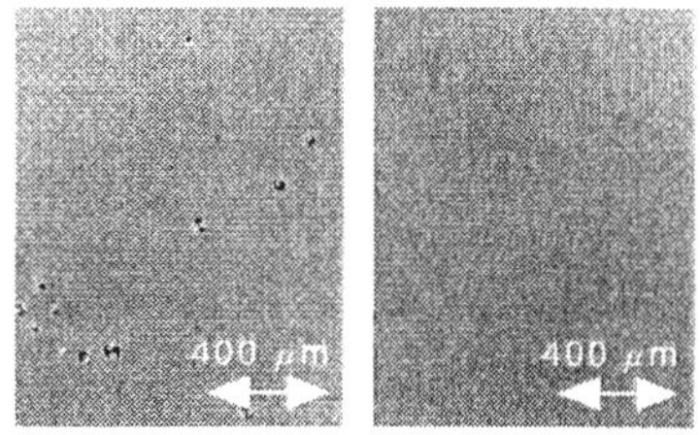

(a) $C_H = 3.1$ at.%. (b) $C_H = 1.1$ at.%.

Fig.4 Surface morphology for films with different H contents (C_R).

of the poly-Si films after RTA at 1000°C for only 5 s. a-Si films before RTA have H contents of 3.1 at.% and 1.1 at.%, respectively. No bubble exists on the surface of the film at 1.1 at.%, although some bubbles exist on the film at 3.1 at.%. It obviously shows that low H contents surppress the creation of bubbles and keep flat surfaces.

Crystallinity and film structure

Figure 5 shows Raman spectra for Cat-CVD Si films before and after RTA. By annealing at 1000°C for only 5 s, the films are crystallized and weak intensity of Raman signal for a-Si phase is observed. In any case, the peak positions for the signal due to the crystalline phase are located at 518-519 cm^{-1} and the values are somewhat smaller than those for the single-crystalline Si at 520 cm^{-1}. However, the difference is not significant and the tensile stress is very small particularly compared with ELA poly-Si films, in which the peaks are located at approximately 515-517 cm^{-1} [3] due to large tensile stress.

Structure of the poly-Si films by RTA

Figure 6 shows optical microscopic photographs of surface of the poly-Si films, after RTA at 1000°C for 5 s, and that after FA at 1000°C for 30 min, respectively. The surface after RTA is much smoother than that after FA. It is also known that the annealing at high temperatures

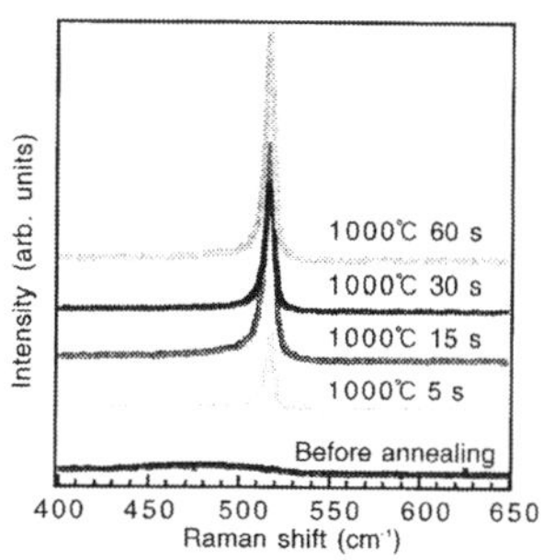

Fig.5 Raman spectra for the Si films before and after RTA.

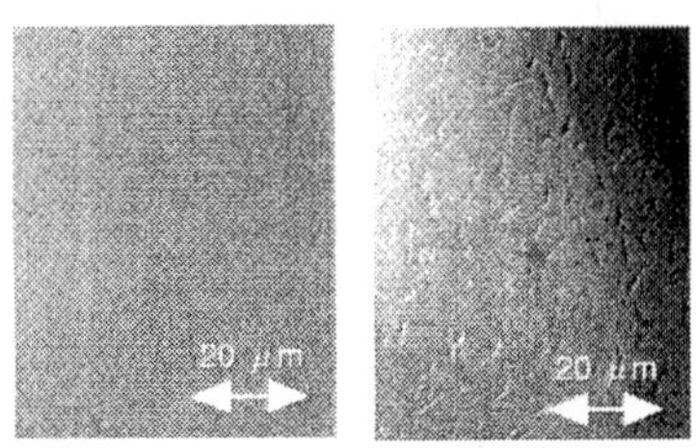

(a)1000℃, 5 s (b)1000℃, 30 min

Fig.6 Optical microscope images of the surface for poly-Si films after annealing by RTA at 1000°C for 5s (a) and in a furnace at 1000°C for 30 min (b).

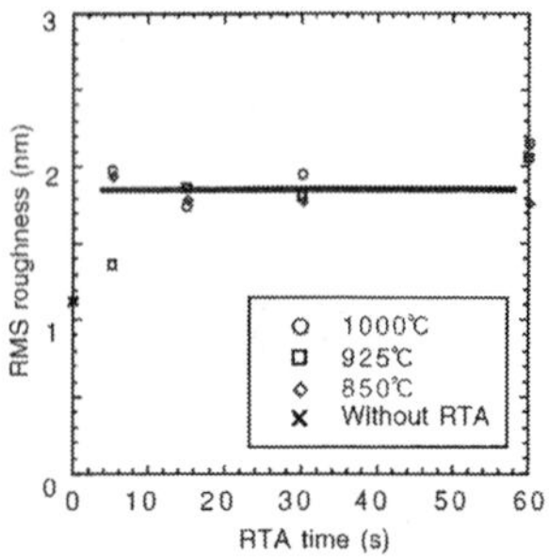

Fig.7 Surface RMS roughness for poly-Si films formed by RTA as a function of RTA time.

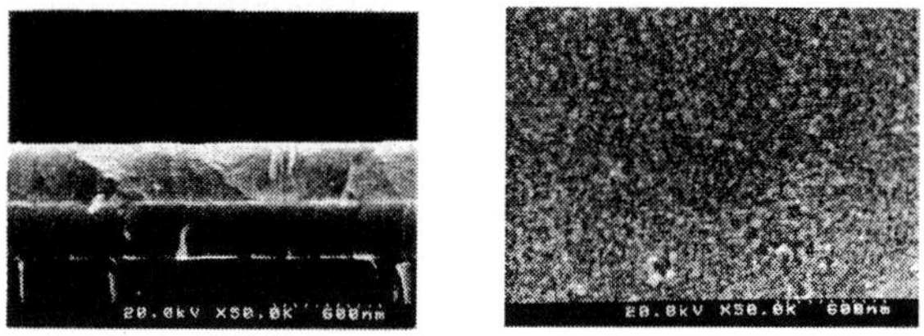

Fig.8 Cross-sectional SEM view for poly-Si films after RTA in N_2 at 1000°C for 5 s (a) and the surface SEM view after Secco-etch treament (b).

for long time causes rough surface of poly-Si because of crystal growth and increasing of grain size.

Figure 7 shows the root mean square (RMS) roughness, estimated from AFM measurements as a function of annealing time. The scanned area for AFM was 20×20 μm^2. Even after RTA for 60 s, the roughness is only 2.2 nm. Thus, it is clear that flat surface is obtained due to short process-time of RTA.

The structure of the poly-Si films was investigated detailedly by scanning electron microscope (SEM). Figure 8 (a) shows cross-sectional view of the film after RTA at 1000°C for 5 s and (b) shows surface view after Secco-etch treatment to make it easier to observe the grains. It is clearly shown that the films are dense and composed of fine grains. From the photograph, the grain size is evaluated to be about 50 nm. The small surface roughness is probably attributed to the existence of the small grains.

Electrical properties

Figure 9 also shows the resistivity as a function of RTA time. The resistivity significantly

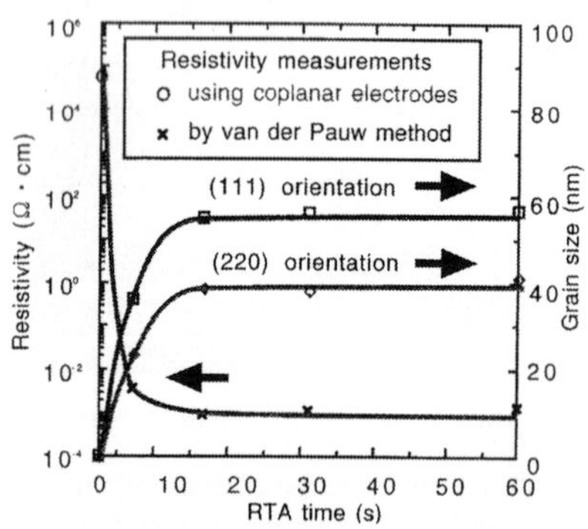

Fig.9 Resistivity and grain size for poly-Si films by RTA as a function of RTA time.

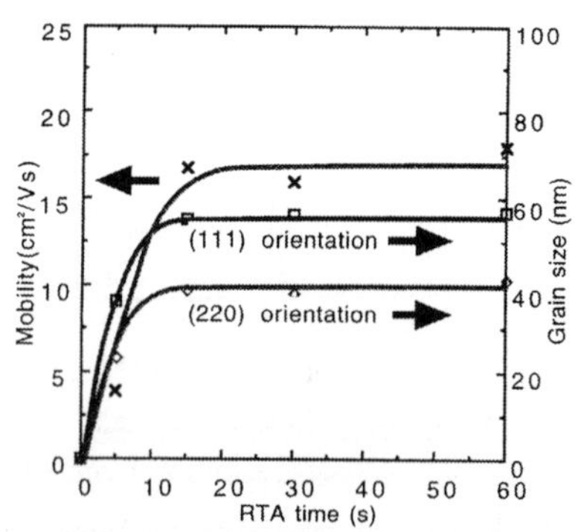

Fig.10 Mobility and grain size for poly-Si films by RTA as a function of RTA time.

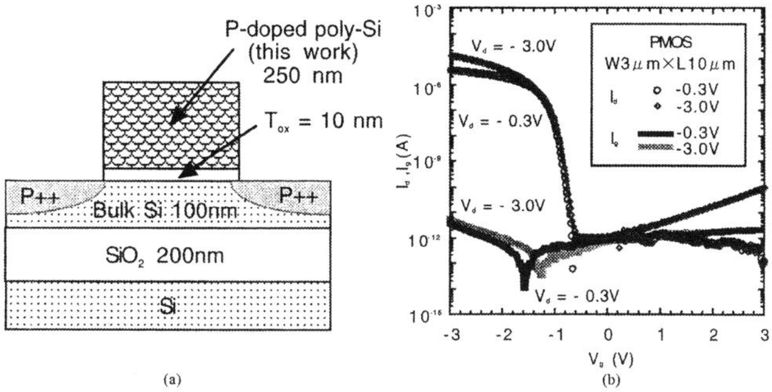

Fig.11 Structure (a) and V_g-I_d and V_g-I_g characteristics (b) of the fabricated MOSFET with the poly-Si film as gate electrode by RTA at 1000°C for 5 s.

decreases to 0.004 Ω·cm by the RTA for only 5 s and it is saturated at approximately 0.001 Ω·cm. Figure 10 shows the carrier mobility obtained using the van der Pauw method. The carrier concentration is 4.4×10^{20} cm^{-3} by the annealing for 5 s and almost all P atoms are activated by the RTA. The mobility is also likely to depend on the grain size. The saturation of both resistivity and mobility occurs after annealing for the duration longer than 15 s and the tendency is similar to that of grain size. These phenomena appear to imply that the rapid reduction of the resistivity after annealing is not only related to the impurity activation but also to the enlargement of grain size.

Finally, to investigate the feasibility of their device application, MOSFETs were fabricated on single-crystal Si with the present poly-Si gate electrode as shown in Fig. 11(a), and the V_g-I_d and V_g-I_g Properties are shown in Fig. 11(b). The MOSFET operates normally and ON/OFF ratio reached over 10^8. The gate-leakage current density is 1.8×10^{-7} A/cm^2 at the gate voltage of -5.0 V and serious leakage is not observed. The C-V profiles showed no hysteresis loop. According to the results, it is clear that the crystallization of a-Si films by RTA does not induce serious damage in the thin gate oxide. The thickness of oxide estimated by the C-V measurements was 12.0 nm. Similar measurements were carried out using aluminum electrodes instead of the poly-Si films. In this case, the thickness was 11.2 nm. This means that the poly-Si films are slightly depleted up to only a 0.8 nm-depth in equivalent-oxide value. However, since the value is very small and negligible, the poly-Si gate electrodes are feasible for use in MOS gate electrodes.

4.Conclusions

In the present work, the following results are obtained;

(1) P-doped Cat-CVD a-Si films with low H content at approximately 1 at.% are suitable for solid-phase crystallization (SPC) process.

(2) RTA is effective to keep surface smoothness.

(3) High crystallinity, smooth flatness around 2-nm-RMS roughness and low resistivity of 0.004 Ω·cm are all obtained using the RTA treatments of Cat-CVD a-Si films at

1000°C only for 5 s.

Acknowledgements

The authors would like to thank Dr. Susumu Horita for useful discussions and also thank Mr. Takayuki Tsutsumi and Ms. Chisato Yokomori for kind help of experiments.

References

[1] A. H. Mahan, J. Carapella, B. P. Nelson, R. S. Crandall and I. Balberg: J. Appl. Phys. **69** (1991) 6728.

[2] A. A. Langford, M. L. Fleet, B. P. Nelson, W. A. Lanford and N. Maley: Phys. Rev. B **45** (1992) 13367.

[3] A. Hara and M. Okabe: Fujitsu **49** (1998) 191 [in Japanese].

Rapid Thermal Processing for Future Semiconductor Devices
H. Fukuda (Editor)

Evaluation of Crystalline Defects in Thin, Strained Silicon-Germanium Epitaxial Layers by Optical Shallow Defect Analyzer

Yukihiro Kiyota and Kazuo Takeda
Central Research Laboratory, Hitachi, Ltd.
1-280 Higashi-Koigakubo, Kokubunji, Tokyo, 185-8601, Japan
Phone: +81-42-323-1111 Fax: +81-42-323-7764 E-mail: kiyota@crl.hitachi.co.jp

1.Introduction

As the performance of silicon transistors is improved by lateral and vertical scaling, there is an increasing demand for thin epitaxial silicon or silicon-germanium layers in ULSI fabrication. Applications of epitaxial layers include the base region of the hetero-junction bipolar transistor (HBT), the channel region of complementary metal oxide semiconductor (CMOS) transistors, and the elevated source and drain regions of CMOS transistors. Among these devices, HBTs are already fabricated on production lines and many kinds of LSIs using silicon-germanium HBTs are commercially available. The silicon-germanium HBT has great potential for applications in high-speed telecommunication systems and wireless communication systems [1, 2]. Silicon-germanium epitaxial growth, which is a key process in HBT fabrication, has reached mass-production level [2, 3], however, a method to evaluate the process quality, especially one for evaluating the crystallinity of the grown layers, has not yet been established.

The crystallinity of silicon epitaxial layers is usually evaluated by defect etching, lifetime measurement, and so on. These methods can be used to evaluate thick (more than a few microns) epitaxial layers. However, in recent applications of epitaxial layers in silicon LSIs, the thickness of the epitaxial layers is below 100 nm. Hence, new analytical techniques for evaluating the crystallinity of thin epitaxial layers are needed.

To meet the above-mentioned requirement, we have developed a non-destructive method for detecting defects in the near-surface region of a silicon substrate [4, 5]. The equipment, called an optical shallow defect analyzer (OSDA) estimates the size and depth of the defects by measuring two scattered light beams from each defects at two wavelengths. In this study, we used the OSDA to evaluate the crystallinity of thin strained silicon-germanium layers.

2.Experiments

Silicon-germanium Epitaxial Growth

Silicon-germanium epitaxial layers were grown by one-wafer-type ultra-high-vacuum chemical vapor deposition (UHVCVD). The experimental conditions are listed in Table 1.

Table.1 Process conditions

Cleaning		H2 annealing	Si2H6 gas etching
	Temp.	750°C	
	Flow rate	H2: 4 L/min	Si2H6: 2 ml/min
	Pressure	1300 Pa	< 1 Pa
	Time	10 min	15 s
SiGe growth	Temp.	550°C	
	Flow rate	Si2H6: 2 ml/min	
		GeH4: 0 - 4.8 ml/min	
	Ge content	0, 5, 10, 15, 20%	
	Growth time	6 - 55 min	
	Thickness	50 nm	

Two methods for cleaning the wafer surface were used: hydrogen annealing and Si_2H_6 gas etching [6]. Hydrogen annealing is a popular method for conventional atmospheric or low-pressure epitaxial growth and is also effective for UHVCVD [7]. Fifty-nanometer-thick silicon-germanium layers with various germanium contents were grown on 8-inch silicon epitaxial wafers. Epitaxial wafers are suitable for separating the defects in thin epitaxial layers from those in the substrate since they include a much lower number of defects than those in Czochralski (CZ) wafers [5].

Optical Shallow Defect Analyzer

The optical shallow defect analyzer (OSDA) is schematically shown in Fig. 1. The incident lights, each with a different wavelength (532 or 810 nm), are scattered from each defect in the silicon wafer and are detected by two photomultipliers. Silicon has different absorption

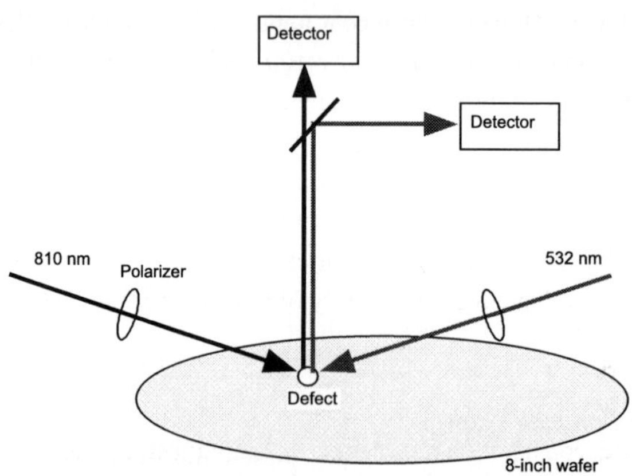

Fig.1 OSDA measurement system.

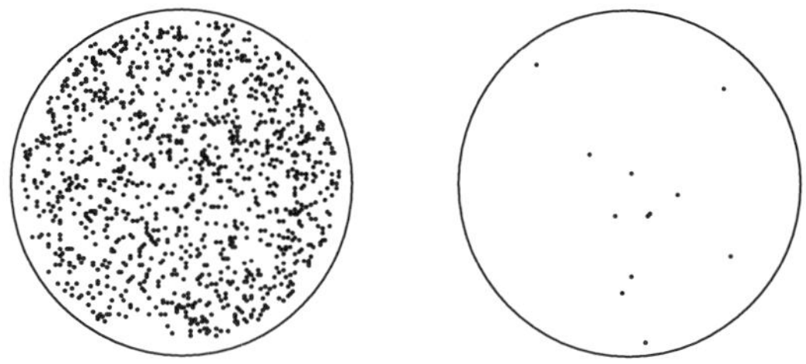

(a) CZ wafer: 19.1 defects/cm^2 (b) Epi wafer: 0.2 defects/cm^2

Fig.2 Evaleation of defects in an epitaxial water and Czochralski water.Each dot represents detected defect.

coefficients (k) at the two wavelengths (k=0.045 at 532 nm, 0.006 at 810 nm). The penetration depths where the intensity of lights decays down to 1/e are 2 m for the 532-nm wavelength and 20 m for the 810 nm. The intensity of two scattered lights beams from a defect decreases exponentially with the depth of the defect according to the Lambert-Beer theory, and it is proportional to the 6th power of the defect size according to Rayleigh scattering theory. Depth Z and size d of the defect are determined from these theories with the following equations [4, 5].

$$Z = C_1 \ln\left(\frac{I_{810}}{I_{532}}\right) + C_2 \qquad (1)$$

$$\ln(d) = \left(\frac{1}{6}\right)\ln(I_{810}) + C_3 \qquad (2)$$

Where, I_{810} and I_{532} are the intensities of the scattered lights at 810 nm and 532 nm, and C_1, C_2, and C_3 are constants that depend on the OSDA. To quantify the defect size, polystyrene particles with known size were used as a standard. (This is a well-known method used in a particle counter on an LSI production line.)

3.Results and discussions

Wafer evaluation

First, the defect densities in the CZ and epitaxial wafers were evaluated by the OSDA. The structure of the epitaxial wafer consisted of a 3- m p-type epitaxial layer on a p-type substrate. A lot of defects were detected from the CZ wafers. These defects are grown-in defects generated during the pulling up of the silicon ingot [4, 5]. In contrast, the epitaxial wafer included a small number of defects, indicating that defects in a thin epitaxial layer can be analyzed by using epitaxial wafers because the number of defects in a wafer can be neglected.

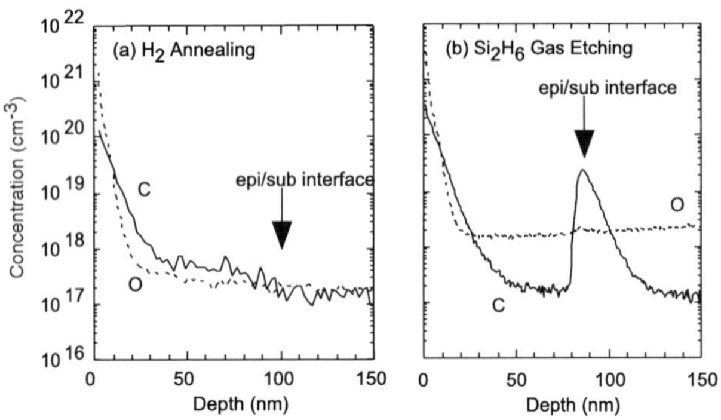

Fig.3 SIMS profiles.

Effect of Wafer Surface Cleaning

Contamination levels at the interface between the epitaxial layer and the substrate of samples cleaned by different surface-cleaning methods were evaluated by secondary ion mass spectroscopy (SIMS) and are shown in Fig. 3. It is clear that hydrogen annealing effectively reduces carbon and oxygen concentrations at the interface to below the detection limit of SIMS; however, carbon contamination could not be removed by Si_2H_6 gas etching. The OSDA defect maps of the samples are shown in Fig. 4. The thickness of the samples was 50 nm and germanium content was 15%. Each dot represents a defect. The defects detected in the hydrogen-annealed sample are probably particles that were mixed with the film during growth. In the case of the Si_2H_6-gas-etched sample, we could not measure the whole wafer because there were too many defects. Because of the lack of depth resolution of the OSDA, we cannot conclude that these defects arose from carbon contamination at the interface. However, we believe this contamination is the main reason for the many defects. The difference between the crystallinities of the two samples shown in Fig. 4 could not be identified by other analytical methods such as the transmission electron microscope (TEM) or Rutherford backscattering spectroscopy (RBS). It is clear that OSDA can detect defects that cannot be detected by other methods and that it can also detect defects in thin epitaxial layers.

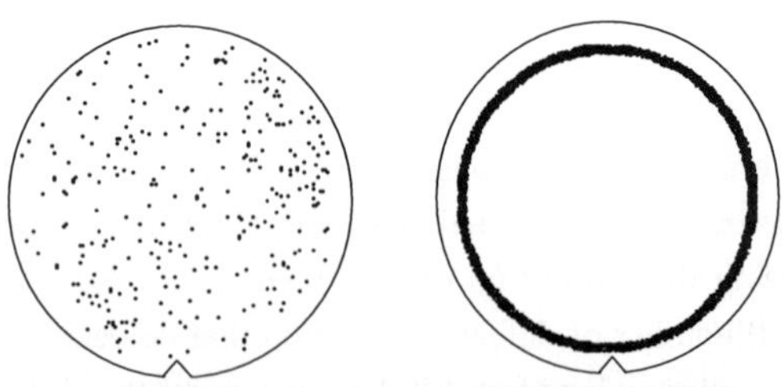

(a) H_2 annealing. Si_{85} Ge_{15}
9.1 defects/ cm^2

(b) Si_2 H_6 gas etching,Si_{85} Ge_{15}
1939 defects/cm^2
(only data at peripheral are shown)

Fig.4 OSDA defects

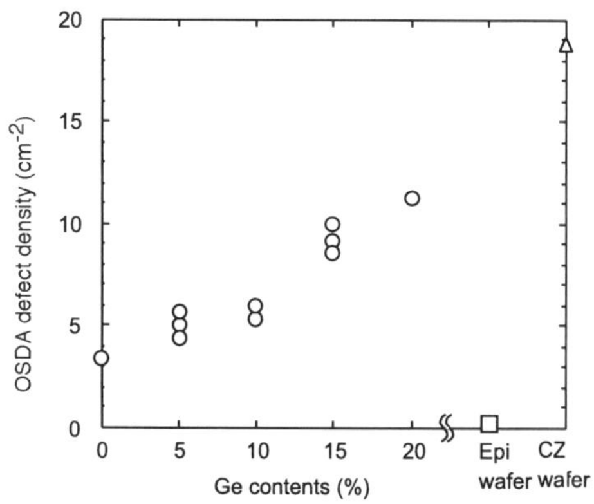

Fig.5 OSDA defect density vs. Ge content.

Effect of Germanium Content

Fifty-nanometer-thick silicon-germanium epitaxial layers with various germanium contents were grown on an epitaxial wafer following hydrogen annealing, and their defect density was analyzed by OSDA (Fig. 5). Samples with germanium concentrations of 5, 10, and 15%, were grown and measured several times in order to evaluate the measurement accuracy. The variations are fairly small as shown in the figure so the accuracy of the data is considered to be sufficient. It is clear from the figure that the density of the defects increases the germanium concentration increases. Moreover, TEM photographs showed that there were no dislocations caused by strain relaxation in the samples. We thus conclude that the detected defects must be particles, clusters of vacancies, or other phenomena. Further study will, however, be needed to identify them precisely.

4.Summary

An optical shallow defect analyzer was used to measure the density of defects in thin strained silicon-germanium epitaxial layers. Its sensitivity was sufficient to clarify the difference in the crystallinities of samples cleaned by different surface-cleaning methods. The OSDA showed that the defect density increases with increasing germanium concentration.

Acknowledgments

The authors express their sincere thanks to Yasushi Matsuda of the Semiconductor group, Hitachi, Ltd. for his continuous support and encouragement. They also thank Katsuya Oda of the Central Research Laboratory, Hitachi, Ltd. and Professor Taro Inada of Hosei University for their useful discussions on the characteristics of silicon-germanium epitaxial layers.

References

[1] K. Washio, E. Ohue, H. Shimamoto, K. Oda, R. Hayami, Y. Kiyota, M. Tanabe, T. Hashimoto, and T. Harada, Tech. Digest of International Electron Device Meetings, p. 741, (2000).

[2] Y. Kiyota, T. Udo, T. Hashimoto, A. Kodama, H. Shimamoto, R. Hayami, E. Ohue, and K. Washio, Extended Abstracts of International Conference on Solid State Devices and Materials, p. 372, (2001).

[3] K. Wolf, W. Klein, N. Elbel, A. Berthold, S. Grondahl, T. Huttner, S. Drexl, and R. Lachner, IEICE Trans. Electron., E84-C, p.1399, (2001).

[4] K. Takeda, H. Ishida, and A. Hiraiwa, Extended Abstracts of International Conference on Solid State Devices and Materials, p. 151, (1996).

[5] K. Takeda, Hyomen Kagaku, 22, p. 323, (2001) (in Japanese).

[6] H. Hirayama, T. Tatsumi, A. Ogura, and N. Aizaki, Appl. Phys. Lett., 51, p. 2213, (1987).

[7] K. Oda and Y. Kiyota, J. Electrochem. Soc., 143, p. 2361, (1996).

Rapid Thermal Processing for Future Semiconductor Devices
H. Fukuda (Editor)

Novel UV-assisted Rapid Thermal Annealing of Ferroelectric Materials

E. Lynch, S. O'Brien,P.V. Kelly
NMRC, Lee Maltings, Prospect Row, Cork; Ireland
H. Guillon
JIPELEC. 11, Chemin du Vieux Chene, F-38240 Meylan
L.Pardo, R. Poyato, A. González, M.L. Calzada,
Insto. Ciencia de Materiales de Madrid-CSIC. Cantoblanco. E-28049 Madrid
Igor Stolichnov
Ceramics Laboratory, EPFL, Swiss Federal Institute of Technology, CH-1015 Lausanne

We report the crystallisation of the ferroelectric perovskite phase of modified lead titanate (mPT) from a sol-gel layer at a maximum temperature of 550°C using ultraviolet-assisted rapid thermal processing (UV-RTP). This compares well to the temperature of 650°C required to develop the perovskite phase of modified lead titanate using purely thermal rapid thermal processing. The reduction in the maximum processing temperature achieved by UV-assisted rapid thermal processing opens the possibility for wider applications of ferroelectrics on thermally sensitive substrates and structures.

1.Introduction

This paper describes a novel processing technique termed UV-RTP, which is used for reduced temperature crystallisation of ferroelectric thin-films. These materials are attracting significant research interest in applications such as pyroelectric sensors and micro-electro-mechanical systems due to their superior properties. Furnaces and Rapid Thermal Processing (RTP) units find a wide range of application in semi-conductor manufacturing, for example oxidation, nitridation, annealing and glass-reflow processing. Although conventional furnace technology has been established for much longer than RTP technology, RTP has a number of attractive features, which offer advantages over furnace processing. For example, there is no time delay in heating and cooling in RTP. Also the lower thermal budget of an RTP system means that even a single wafer system can have competitive throughput.

It is the associated rapid temperature ramp that is of most interest in the area of ferroelectric materials development. Ferroelectrics are materials that have a spontaneous electric polarisation, the direction of which can be reversed by the application of an external electric field. They display dielectric properties, along with piezoelectric (electrical polarisation related to mechanical stress) and pyroelectric (spontaneous electric polarisation which is temperature dependant) properties. Consequently, ferroelectric thin films find use in a wide range of applications, such as non-volatile-memories, detectors, integrated capacitors and optical waveguides.

Ferroelectric thin film fabrication involves the following steps:
- Precursor selection and preparation
- Deposition of sol-gel onto substrate
- Crystallisation of single phase thin film

The majority of ferroelectric materials possess a perovskite crystal structure seen here in figure 1:

The spontaneous electric polarisation associated with ferroelectric materials occurs as a result of a small lattice deformation of the perovskite structure below the Curie temperature (T_c). Below this, ferroelectric materials have a characteristic hysteresis loop [P vs E plot], but above this temperature are termed paraelectrics.

However, the formation of single-phase perovskite material from an amorphous precursor occurs via a transient, meta-stable pyrochlore phase. This pyrochlore, if stabilised, is non-ferroelectric and its presence in the film will limit the overall ferroelectric properties. This transient phase occurs in the temperature range 350°C–400°C or during nucleation. To prevent stabilisation of this phase, prolonged time at the temperature range 400°C–500°C is avoided. This is where the use of RTP has been most successful. In practice, it has been found that RTP of deposited precursor materials is the most efficient means of obtaining single-phase material, as the temperature is ramped through the 350°C–450°C range, typically at a rate of ~50°C s^{-1}.

When an external electric field is applied to an untreated ferroelectric material, all domains tend to be polarized along the applied field direction. At sufficient field, an almost single domain state is effectively achieved. This is termed saturation polarization P_s. When the applied field is removed, there is residual alignment to the original applied field direction, termed remanent polarization (P_r). At a certain voltage, termed the coercive field, P_r can be reversed by application of a reverse field, which can be further increased with greater voltage.

In spite of recent advances in the area of ferroelectric thin film technology, it is still necessary to further develop and optimise the processing involved. The major concerns in industrial thin

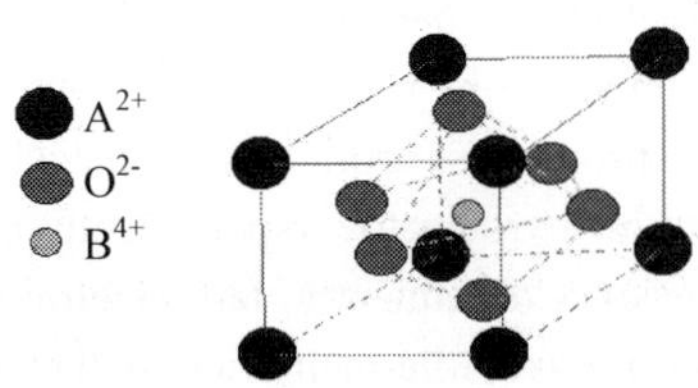

Fig.1 Perovskite Crystal Structure. (Face centred cubic)

film fabrication for a ferroelectric manufacturing process are related to stoichiometry and uniformity control, reproducibility and deposition efficiency. Outside of these concerns, the biggest issue associated with the wide-scale use of ferroelectric materials is the high processing temperature. For a standard ferroelectric process, it is impossible to reduce the processing temperature to below 650°C and in some cases, the temperature has to be increased to 800°C to give the correct crystal phase. This limits the potential substrates as the development process of the films damages all temperature-sensitive substrates. Also, the incorporation of CMOS-type structures under the film is prevented due to thermal damage of the devices above 470°C. This is the singular issue which has limited the potential applications of the ferroelectric technology.

Recently, a new generation of ultra-violet lamps has been developed which are capable of producing high-power radiation from the near UV (354nm) to the vacuum UV (108nm). The operation of these lamps relies on the radiative decomposition of excimer states created by a barrier discharge. The development of these low-cost, high power and large-area UV and VUV sources promises enormous potential for materials' processing and several applications have already been demonstrated. The use of a variety of excimer lamps for low temperature deposition of metals or metal oxides via decomposition of spin-on coatings of metalorganic precursors or sol gel solution has been demonstrated. It has also been demonstrated that UV annealing can significantly improve the properties of these films.

Unlike most other lamps, excimer UV sources provide intense UV radiation at specific wavelengths. This narrow-band UV radiation can initiate chemical reactions, break molecular bonds or modify surface properties. Like other UV sources, excimer lamps can also be used to induce photo-polymerisation of certain monomer or oligomer and sol-gel systems.

Due to the benefits of UV excimer lamp processing of thin-film deposition, it was decided to investigate the effect of UV processing in sol-gel derived ferroelectric thin film formation. A sol is a colloidal suspension of solid particles in a liquid. In sol-gel processes, precursors consist of a metal or metalloid element surrounded by various ligands, i.e. organic groups. Sol-gel technology is attracting immense attention at present due to its easy doping, superior functionality and low-cost mass-production, particularly in the area of stoichiometrically controlled mixed oxide production.

The use of UV technology in processes such as cross-linking of polymers etc., is very different to the application proposed in this work. The formation of the perovskite crystal phase cannot be achieved by means of UV-irradiation alone. A thermal process is required to initiate nucleation and crystallisation of this crystal phase from the amorphous dried sol-gel precursor. The combination of the standard RTP process with these UV lamp systems formed the basis for this work. In general, temperatures of 650°C are necessary for the formation of quality single-phase perovskite material from a suitable sol-gel precursor system. With this technology, the processing temperature is significantly reduced with no forfeit in film quality.

2.Experimental

UV-RTP processing of these samples was performed in the "JIPELEC JetClip sg" reactor. The design of a conventional furnace was altered to allow the incorporation of the UV lamp system along with the IR system. The UV-RTP chamber design can be seen in figure 2, below:

The chamber consists of two cooled KrCl* high-power excimer lamps on the upper quadrant and a standard 24 lamp IR system on the lower surface. These IR lamps are capable of heating from ambient temperatures to 1000°C at a rate of at least 50°C/s. The environment within the chamber is controlled by means of mass-flow controllers and vacuum pumps. Processing is controlled by a PC, which also allows monitoring of the temperature and UV power during processing.

For the purposes of this study, a calcium modified lead titanate sol-gel system was used as a precursor for the ferroelectric thin-film formation. The modified lead titanate thin films were developed from a previously reported sol-gel solution, with the nominal compositions of $Pb_{0.76}Ca_{0.24}TiO_3$[6]. This was deposited on a special $Si/SiO_2/TiO_2/Pt/TiO_2$ substrate at 2000rpm and pyrolysed at 150°C-250°C for 5 minutes.
A variety of test processes were performed in the JetClip sg UV-RTP system to determine the conditions required to give high quality ferroelectric material, using UV irradiation and reduced processing temperatures.

Structural analysis was performed by glancing angle XRD. Electrical characterisation was then performed on single-phase perovskite samples. This consisted of measurement of the dielectric constant and loss followed by the pyroelectric coefficient for these films. This was used as a measure of the overall properties of the films, given that the pyroelectric values are

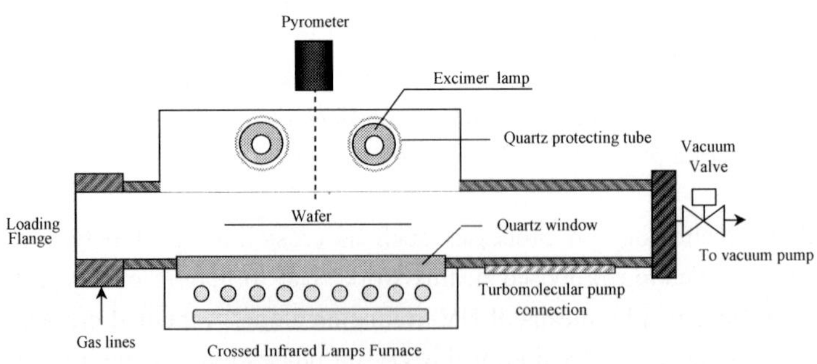

Fig.2 UV-RTP furnace.

indicators of the polarisation perpendicular to the surface. Further to this the polarisation of the samples and piezoelectric values were also measured.

3.Results and Discussion

Crystallisation of single-phase perovskite from a dried sol-gel precursor system usually requires thermal energy for the nucleation and growth of the crystal phase. Using conventional processes, a maximum temperature of 650°C is required for modified lead titanate thin films. This temperature is higher for other systems such as strontium bismuth tantalate, which requires temperature in the order of 800°C.

Samples were processed by the standard 650°C RTP cycle in the UV-RTP system to assess the properties developed at a temperature of 650°C, seen in figure 3:

The calcium-modified lead titanate (Ca-mPT) film prepared according to this recipe was found to have a dielectric constant of ~160, dielectric loss of less than 0.06 and a pyroelectric coefficient of -24.9×10^{-9} C cm^{-2} K^{-1}. These are the basic result which would be desirable for any reduced temperature film developed by UV-RTP

Further parameters commonly measured for these films are the piezoelectric coefficients. The texture and orientation of a thin film will affect the piezoelectric measurements, with a random orientation giving a zero piezoelectricity. The $d_{33,}$ otherwise known as the longitudinal coefficient, can only occur along the polar direction and develops parallel to an applied stress. It is found to be high for (111) oriented films after polarisation and here has a value of 40 pm/V. Formation of the (111) type film gives a value of 0.4 C cm^{-2}.

Initial UV-RTP studied the ability of the UV exposure to enhance the growth and development of the crystal phase. The recipe used a maximum temperature of 550°C, and incorporated the UV exposure at this temperature. The UV power is written in figure 4 as 75% UV power, a

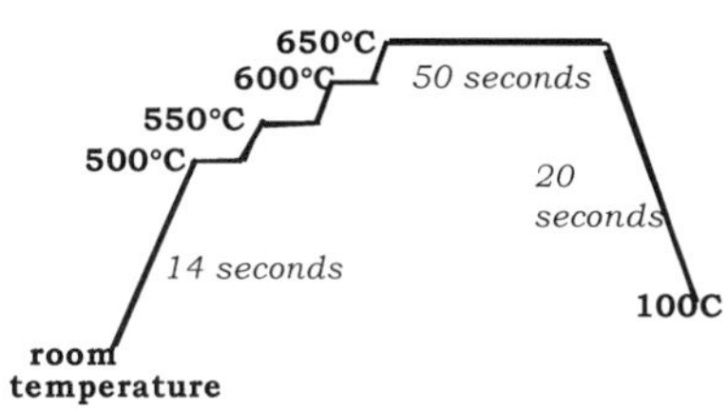

Fig.3 The RTP process requires rapid heating to 500°C and further annealing to 650°C for crystal phase development.

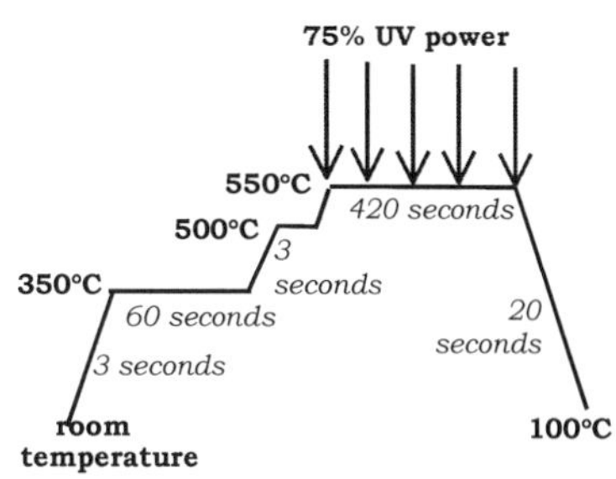

Fig.4 UV-RTP recipe for process 1, with maximum processing temperature and UV irradiation at 550°C.

notation indicating the75% of the full power of the lamps was used (the full power is considered as 120mW/cm^2, so 75% is still 90mW/cm^2)

XRD analysis was completed for the Ca-mPT prepared by UV-RTP according to the recipe described above. The XRD data shown in figure 5 indicates that the material produced by this process incorporates the non-ferroelectric, pyrochlore crystal phase.

As expected, when electrical characterisation was performed, the samples were found to have poor electrical properties. The dielectric constant was very low ,ε' = 100, while the dielectric loss was too high, tanδ = 0.12. The pyroelectric coefficient was found to be 7.5 x 10^{-9}C cm^{-2} K^{-1}, which leads to the conclusion that UV irradiation at 550°C will not give suitable film quality.

With the incorporation of the UV irradiation at a lower temperature, as depicted in figure 6, complete precursor decomposition is facilitated.

Samples prepared by this process were measured by XRD and proved to be single phase, perovskite materials with a mixed orientation. The dielectric constant was 166 while the loss tangent was 0.08, both values comparing well to the RTP sample mentioned previously. The maximum pyroelectric coefficient was 16 x 10^{-9} Ccm^{-2}K^{-1}, comparable to the RTP result and suitable for most pyroelectric applications.
A comparison of the polarisation curves was recorded for both the RTP sample and sample from the process in figure 6.

From this graph, the remnant polarisation values (Pr) were then obtained. P_r for the RTP sample was recorded at28μC cm^{-2}. While UV-RTP process 2 resulted in a value of 22 μC cm^{-2}. The square, centred loop of the RTP sample is indicative of a good quality material with rapid reversal. The thin loop means that the required voltage for saturation is lower for the RTP

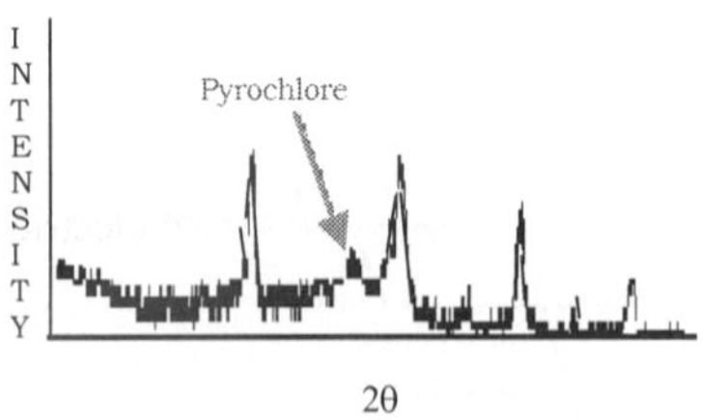

Fig.5 XRD data showing secondary phase development.

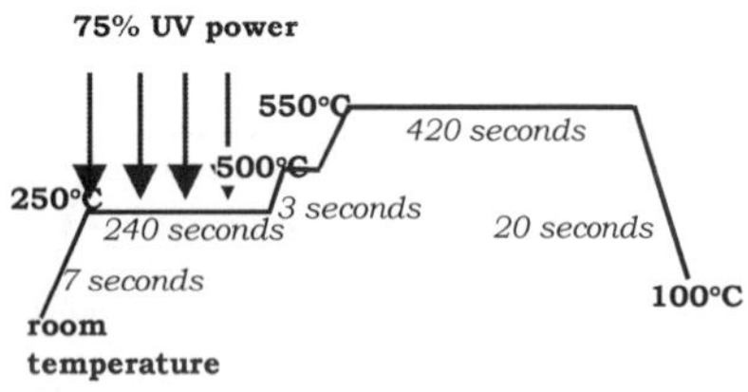

Fig.6 UV-RTP process 2.
UV is used for bond cleavage, while the thermal process at 550°C is used for development of the perovskite crystal phase.

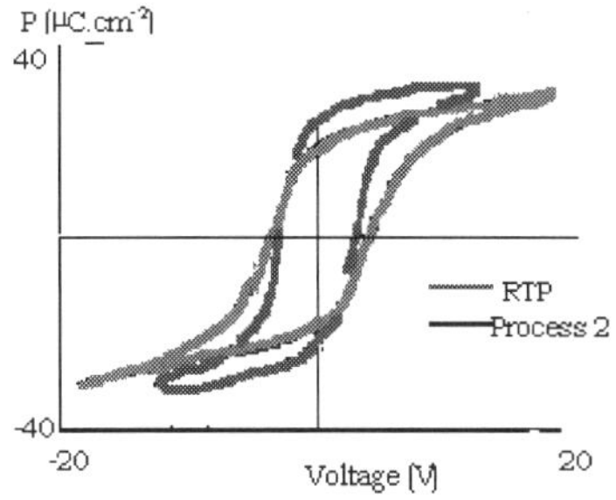

Fig.7 Polarisation data collected for Ca-mPT films prepared by RTP (650°C) and UV-RTP process 2 (550°C).

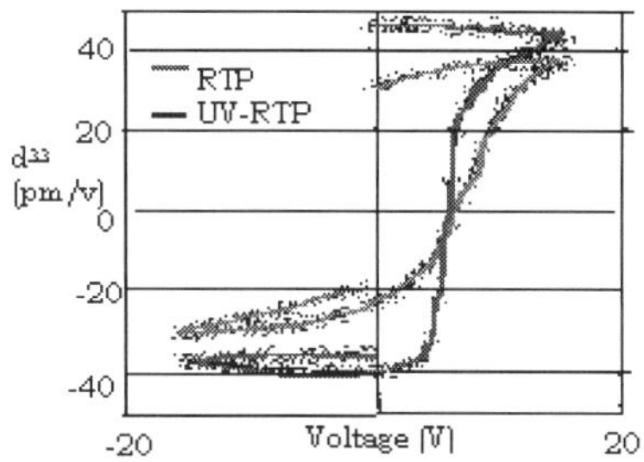

Fig.8 d_{33} data collected from Ca-mPT prepared by standard RTP and UV-RTP process no. 2.

sample in this case, while the reversal of the UV-RTP is marginally slower than for the RTP sample. Overall, however, the results compared well for the sample processed at 550°C.

Piezoelectric properties, in this case d_{33}, were then performed on the same samples. The data obtained are shown in figure 8, with the RTP giving the slightly higher value of 45pm/V.

The UV-RTP sample gives a value of 38pm/V, which is still significant. This result, in combination with all the other results indicates that processing at 550°C can achieve ferroelectric properties comparable to those achieved at 650°C, through the use of UV-RTP.

This recipe would seem to be worth further analysis, particularly in the reduction of the processing temperature to 470°C, the maximum temperature for incorporation of these films on CMOS device structures.

4.Conclusion

Good quality calcium-modified lead titanate was prepared at reduced temperature of 550°C by a novel UV-RTP method. Single-phase perovskite with good electrical properties was produced. It was found that UV irradiation performed only at a high temperature was not sufficient to cause the formation of single-crystal perovskite material. A low temperature UV exposure combined with a moderate temperature step gave place to ferro-piezoelectric films, whose properties are indicative of their feasibility to be used in piezoelectric applications.

Acknowledgements

This work was funded by projects BRPR-CT98-0777 (MUVAST) of the EC. Authors wish to thank Ralph Neuberger at EADS for the wafers used in this work.

References

[1] J. M. Salzer, *Solid State Technology*, pp.50 (1997)

[2] G. Uozumi, K Kgeyama, T. Atsuki, N. Soyama, H. Uchida and K. Ogi, *Jap. J. Appl. Phys.*, 38, 5350-5353 (1999)

[3] Y. Fujimori, T. Nakamura and H. Takasu, *J. Appl. Phys.* 38, 5346-5349 (1999)

[4] "Electroceramic Thin Films" O. Auciello and R. Ramesh Eds., MRS Bulletin, July 1996, vol.21 nos 6-7

[5] M. L. Calzada, R. Sirera, F. Carmona, *Journal of Sol-gel Science and Technology* 8 pp.729 (1997)

[6] M .E. Lines, A. M. Glass, *Principles and Applications of Ferroelectrics and Related Materials*, oxford university press, pp.140 (1977)

[7] I. W. Boyd, J. Chang, *Microelectronic Reliability* 40 pp.649 (2000)

[8] M. L. Calzada, J. Mendiola, F. Carmona, P. Ramos, R. Sirera, *Materials Research Bulletin* 31 4 pp.413 (1996

[9] D. J. Macauley, P. V. Kelly, G. M. Crean, *Materials research Society Symposium Proceedings*, 445 pp.45 (1997)

Rapid Thermal Annealing of (1-x)Ta_2O_5-$x$$TiO_2$ Thin Films Formed by Metalorganic Decomposition

K. M. A. Salam, H. Konishi, H. Fukuda and S. Nomura

Department of Electrical and Electronic Engineering, Muroran Institute of Technology,
27-1 Mizumoto, Muroran, Hokkaido 050-8585, Japan
Phone: +81 143 465549 Fax: +81 143 465501 E-mail: s1154047@mmm.muroran-it.ac.jp

Rapid thermal annealing (RTA) of (1-x)Ta_2O_5–$x$$TiO_2$ thin films formed on a Si substrate by metalorganic decomposition (MOD) at different temperature, time and ambient have been investigated. RTA time and ambient do not seem to have a strong effect on the film thickness, crystallinity, and insulating properties. As-deposited films were in the amorphous state and were completely transformed to crystalline after post-deposition annealing above 600°C. A thin interfacial SiO_2 layer was confirmed at the interface between the substrate and dielectric layer. Thin films with $0.92Ta_2O_5$–$0.08TiO_2$ (x=0.08) composition at 900°C in the N_2 ambient for 2 min exhibited superior insulating properties. The main reason for the improvement in the insulating properties of Ta_2O_5 could be the charge compensation of excess oxygen by the TiO_2 additive. The measured dielectric constant and dissipation factor at 1 MHz were 16 and 0.016 respectively for films annealed at 900°C. The interface trap density was 4.5×10^{10} $cm^{-2}eV^{-1}$, and flatband voltage was range from –0.09 to –0.39 V. A charge storage density of 42.5 fC/μm^2 was obtained at an applied electric field of 3 MV/cm. The leakage current density was lower than 1×10^{-9} A/cm^2 up to an applied electric field of 3 MV/cm. The Poole-Frenkel conduction dominates the current-voltage characteristics.

1.Introduction

For MOSFETs scaling into the deep submicron, there is a strong need to reduce the gate oxide. Silicon dioxide (SiO_2)-based gate dielectrics raise a number of fundamental problems for further oxide thickness scaling. The most critical ones are in increasing defect density and tunneling current, resulting in reduced dielectric reliability especially enhanced in decreasing oxide thickness to approximately 20Å. Therefore, scaling of the gate oxide must end or an alternate high-dielectric material showing sufficiently low leakage current must be used. A high-dielectric material with low leakage current is necessary for future generations of dynamic random access memory (DRAM) devices. Tantalum oxide (Ta_2O_5) such a material that is compatible with gate insulator and DRAM cell capacitors. Tantalum oxide thin films are attractive for scaled down capacitor in ultralarge scale integrated (ULSI) circuits because of their high dielectric constant, lower leakage current, lower defect density and high breakdown field strength [1-5]. Extremely reliable Ta_2O_5 films with high temperature stability are strongly required for successful integration into ULSI devices. However, the dielectric and insulating properties of Ta_2O_5 films have been reported to be strongly dependent on the fabrication method, nature of substrate and electrode material, and post deposition annealing treatment condition [6-9]. Ta_2O_5 based composites have been studied for improving the dielectric and insulating properties [1-3,10-11]. We have recently shown that the dielectric and insulating properties of Ta_2O_5 can be dramatically increased through an 8% substitution of TiO_2 [12]. The major focus rapid thermal annealing of $(1\text{-}x)Ta_2O_5\text{-}xTiO_2$ thin films research has been to improve the leakage current characteristics of crystalline Ta_2O_5 thin films deposited on Si substrates. Several fabrication methods such as reactive sputtering [7], thermal oxidation [13], plasma-enhanced chemical vapor deposition (PECVD) [9], low-pressure chemical vapor deposition (LPCVD) [14] have been employed to fabricate good quality Ta_2O_5 thin films.

In this paper, we report the rapid thermal annealing of $(1\text{-}x)Ta_2O_5\text{-}xTiO_2$ thin films formed by metalorganic decomposition (MOD) technique. Recently MOD technique has been extensively used in thin film technology. There are several advantages of the MOD technique such as easier composition control, good homogeneity, low processing temperature, low equipment cost, nonvacuum process, longevity, and uniform deposition over a large substrate surface area [15-16]. The post-deposition annealing of the films was carried out in a rapid thermal annealing system. The RTA process has been widely used in the semiconductor industry for various applications [17-18] like oxidation, nitridation, junction formation, and doping. RTA has emerged as a very attractive technique to overcome some limitations of conventional furnace anneals. Among these are: control of doping [19], control of out-diffusion of ultra-shallow junctions [20], control of analog precision resistors and capacitors [21], and inter-conductor dielectric densification and leveling [22]. A very short annealing time and its relative process simplicity of RTA method as compared with conventional furnace annealing are the great advantages. The objective of short processing time is basically the control of dielectric-silicon interface, bulk structure and carrier transport of the thin dielectric films [6,23].

2.Experimental

Substrate materials used in this study were boron-doped (9-12 Ωcm) Czochralski (CZ)-grown, p-type (100) silicon wafers, which were cut into 1.2 cm×1.2 cm squares. These wafers were first rinsed in deionized water and methyl alcohol and then cleaned by a standard RCA method prior to RTA. Thin films of $(1\text{-}x)Ta_2O_5\text{-}xTiO_2$ were formed by metalorganic decomposition technique using pentaethoxy tantalum ($Ta(OC_2H_5)_5$) and tetraethoxy titanium ($Ti(OC_2H_5)_4$) as precursors. The composition x was controlled by varying the volume content. 2-methoxyethanol ($CH_3OCH_2CH_2OH$) was selected as solvents. Then, the MOD solution was directly deposited onto the Silicon wafers by the two-step spin-coating technique at 2000 rpm for 5 s and then 4000 rpm for 20 s. Varying the 2-methoxyethanol content controlled the viscosity of the MOD solution. Adjusting the viscosity of the MOD solution and spin speed controlled the thickness of the films. After spinning onto various substrates, the films were baked in air on a hot plate at 120°C for 10 min in order to remove the volatile matter. Post-deposition rapid thermal annealing of the films was carried out under various process conditions e.g., varying temperature, time, and ambient using rapid thermal processing apparatus. The film thickness and composition were examined by spectroscopic ellipsometry. The crystallinity of the films was evaluated by X-ray diffractometry (XRD) using Cu Kα radiation at 40 kV. The surface morphology and microstructure of the films were analyzed by atomic force microscopy (AFM) and transmission electron microscopy (TEM) respectively. The $(1\text{-}x)Ta_2O_5\text{-}xTiO_2$ films were electrically characterized by applying capacitance-voltage (*C-V*) and current-voltage (*I-V*) measurements. For this purpose, metal-insulator-semiconductor (MIS) capacitors with an Al gate electrodes of 0.5 mm diameter deposited over the area of the films by vacuum evaporation. After gate patterning, the backside of the wafers were etched to expose silicon substrate and metallized with aluminum to reduce the series resistance. Dielectric loss (tanδ), dielectric constant and charge storage density values were measured on MIS capacitors.

3.Results and Discussion

Figure 1 shows the effect of RTA temperature on film thickness. Although RTA time does not seem to have a strong effect on the film thickness, a little bit change in film thickness for RTA temperature and ambient. The film thickness of oxygen ambient is thicker than that of nitrogen ambient and also increased with increasing RTA temperature. It may be the cause for more diffusion during annealing with oxygen ambient.

The as-deposited films were found to be amorphous, and post-deposition annealing was required to develop crystallinity. Figure 2 shows the XRD patterns of the $(1\text{-}x)Ta_2O_5\text{-}xTiO_2$ (x=0.08) films, annealed in an N_2 ambient by the RTA process, as a function of annealing temperature, while the annealing time was kept at 2 min. All the films thickness was about 32 nm. The XRD patterns indicate the peaks originating from (101), (107), (116), (0016) and

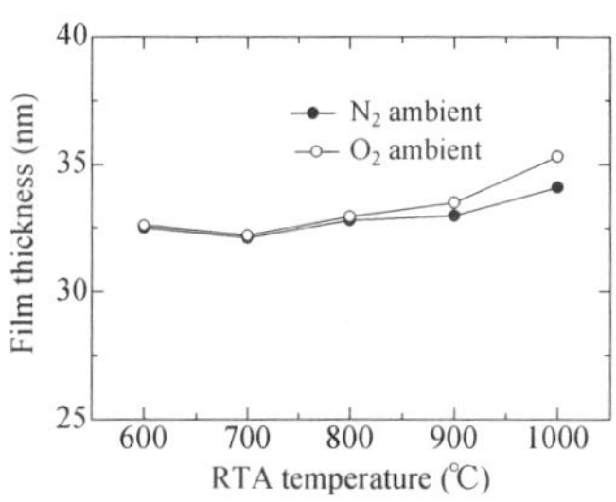

Fig.1 Effect of RTA temperature on the $(1\text{-}x)Ta_2O_5\text{-}xTiO_2$ film thickness at different ambient.

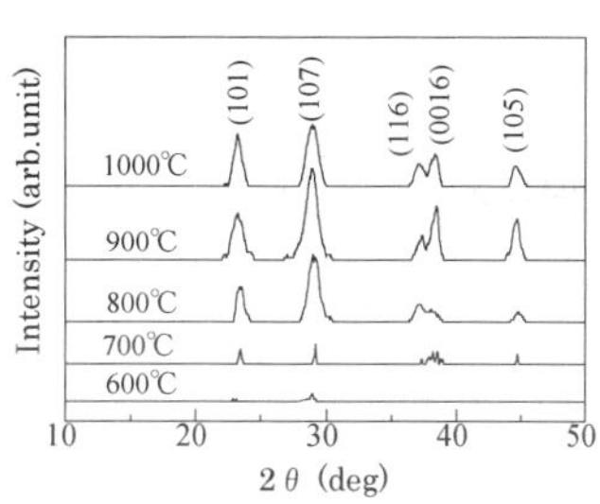

Fig.2 XRD patterns of $0.92Ta_2O_5\text{-}0.08TiO_2$ (x=0.08) films as a function of RTA temperature.

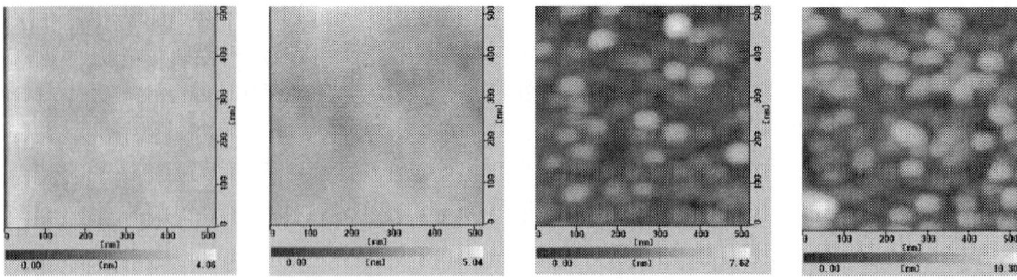

(a) 600°C (b) 700°C (c) 800°C (d) 900°C

Fig.3 AFM photographs of $0.92Ta_2O_5$-$0.08TiO_2$ (x=0.08) films as a function of RTA temperature annealed in N_2 ambient for 2 min.

(105) orientations and was identified to be the β-Ta_2O_5 by JCPDS (No.21-1199) data. Apparently, the peak intensity of (107) orientation had been prominent over other orientations. The XRD data also showed that crystallization occurs above 600°C. As the annealing temperature was increased, the peaks in the XRD pattern became sharper and the full width at half-maximum (FWHM) decreased except (116) orientation indicating better crystallinity and an increase in grain size with increasing annealing temperature. A well-crystallized phase is obtained at an annealing temperature of 900°C. No diffraction peaks in XRD patterns were observed for the films annealed below 600°C indicating that the films were amorphous. During the crystal growth of the film, the reaction starts at the Ta_2O_5/Si interface. After the annealing, the structure changes finally to Ta_2O_5/SiO_2/Si multilayered structure. Above 900°C, an enhancement of SiO_2 layer at the Ta_2O_5/Si interface was observed by spectroscopic ellipsometric measurement. Thus, the Ta_2O_5 film structure seems to depend critically on the presence of underlying SiO_2 layer [24], but RTA time did not seem to have a strong effect on structure and equivalent oxide thickness. In our previous report, we found that the peak intensity of XRD pattern also depends strongly on the composition x and the highest peak intensity was obtained for x=0.08 composition [12].

The effect of the post-deposition RTA temperature on the surface morphology of the

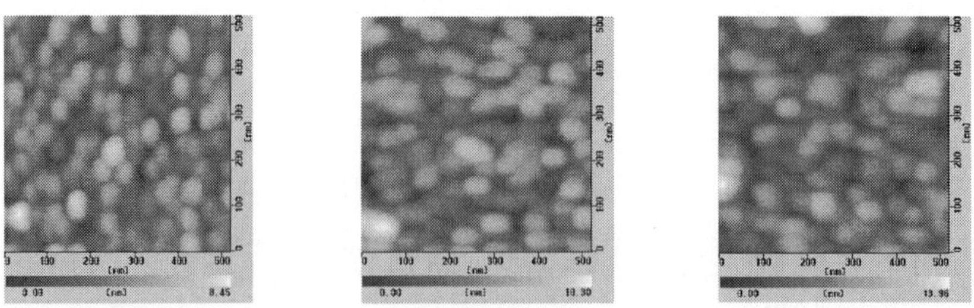

(a) 17 nm (b) 32 nm (c) 50 nm

Fig.4 AFM photographs of (a) 17 nm, (b) 32 nm, and (c) 50 nm thick $0.92Ta_2O_5$-$0.08TiO_2$ (x=0.08) films annealed by RTA process at 900°C in N_2 ambient for 2 min.

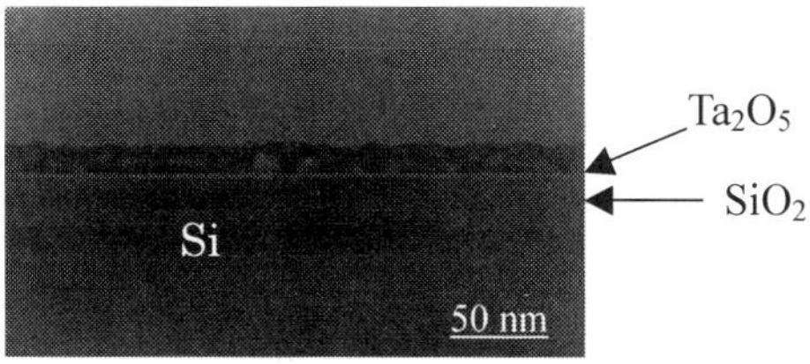

Fig.4 Transmission electron micrograph (TEM) of $0.92Ta_2O_5$-$0.08TiO_2$ (x=0.08)film annealed by RTA process at 900°C in N_2 ambient for 2 min.

$0.92Ta_2O_5$–$0.08TiO_2$ (x=0.08) films are shown in Fig.3. The surface structure of the films was smooth, with no cracks and defects. The films exhibited a dense microstructure and fine grain size. The average surface roughness was found to increase with increase in annealing temperature. The average values of the surface roughness were 4.06nm, 5.04nm, 7.62nm, and 10.30nm at 600°C, 700°C, 800°C, and 900°C respectively. The surface structures of the films were changed with changing annealing temperature. Up to 700°C, the surface structures of the films was smooth with very small grain sizes indicating the film surfaces were not well crystallized with no cracks. Above 700°C, the grain sizes were increased with increasing annealing temperature (Fig.3 (c) & (d)) with a dense microstructure indicating well crystallized which is consistent with the XRD studies indicating an increase in peak intensity with annealing temperature. The average grain size of about 50 nm in diameter for films at the temperature of 900°C. From the technical point of view, larger grain sizes are highly expected with increasing annealing temperature for higher surface mobility thus allowing the films to decrease its total energy by growing larger grains and decreasing its grain boundary area.

The grain size of the films was also found to be thickness dependent as shown in Fig.4. At smaller thicknesses, the grain size was found to be small and it is increased with increasing film thickness indicating the grain growth was not complete. Similar thickness dependence of the grain size has been reported for thin films of other dielectric material [25].

Transmission electron micrographs (TEM) of $0.92Ta_2O_5$-$0.08TiO_2$ films confirm the structure. A typical image displays a very narrow interface layer (white line) between the substrate and dielectric film shown in Fig.5. Above the interfacial SiO_2 layer, a darker and crystallized layer can be observed, which is the tantalum oxide layer.

The *C-V* characteristics of $0.92Ta_2O_5$-$0.08TiO_2$ (x=0.08) films, annealed at various temperatures in N_2 ambient for 2 min by RTA, at a frequency of 1 MHz are shown in Fig.5. With increasing temperature, the *C-V* curves show an increase in the accumulation capacitance. Above 900°C, the capacitance, however, decrease in the *C-V* curve. This might be due to the formation of interfacial SiO_2 layer as shown in Fig.4. The *C-V* curves shifted parallel to the negative bias direction due to the interface trap charge and oxide charge [26]. There was no appreciable change in the capacitance curve for RTA time and ambient. The shift in the C-V curves, called the flatband voltage, V_{FB}, defined by

$$V_{FB} = \phi_{ms} - \frac{Q_f + Q_m + Q_{ot}}{C_{ox}}, \tag{1}$$

where ϕ_{ms}, Q_f, Q_m, Q_{ot} and C_{ox} denote the difference in work function between the metal and the semiconductor, oxide fixed charge, mobile ionic charge, oxide trap charge and capacitance of the oxide respectively. The flatband voltage V_{FB}, in the $0.92Ta_2O_5$–$0.08TiO_2$ film, lies on the slope of the *C-V* curve between the maximum and the inversion region, which is range from -0.09 to -0.39 V. The negative values reveal that the presence of positive charge in the oxide film. The interface trap density (D_{it}) was about 4.5×10^{10} $cm^{-2}eV^{-1}$, which was calculated from the *C-V* data using the Terman method [27] at a frequency of 1 MHz. The charge storage density was calculated from *C-V* characteristics as eq. (2).

$$Q_c = \varepsilon_0 \varepsilon_r E, \tag{2}$$

where ε_0 is the permitivity of free space, ε_r is the dielectric constant of the film, and E is the applied electric field. The charge storage density, Q_c was calculated to be 42.5 fC/μm^2 at 3 MV/cm. The *C-V* characteristics curve of $(1-x)Ta_2O_5$-$xTiO_2$ films with various composition at the temperature of 900°C has also been studied in our previous report [12]. The accumulation capacitance was increased with increasing x, and the maximum capacitance is obtained for the composition of x=0.08. The maximum dielectric constant for the $(1-x)Ta_2O_5$-$xTiO_2$ films was about 16 for x=0.08 composition at 900°C in N_2 ambient for 2 min. The dielectric constant is less than that of Ta_2O_5 films deposited on metal substrates [5]. It is possible that the poor crystallization of $(1-x)Ta_2O_5$-$xTiO_2$ films decreases the dielectric constant is also due to formation of an SiO_2 layer between the silicon and $(1-x)Ta_2O_5$-$xTiO_2$ film.

A dielectric material with frequency and thermal stability is required for the reliability of integrated capacitor applications. Figure 6 shows the dielectric constant and dissipation factor of 32-nm-thick $0.92Ta_2O_5$– $0.08TiO_2$ films annealed at 900°C by RTA as a function of

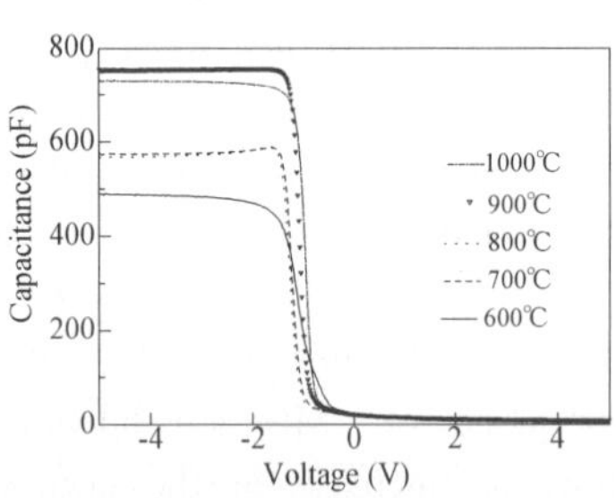

Fig.5 *C-V* characteristics of the $0.92Ta_2O_5$-$0.08TiO_2$ film as a function of annealing temperature.

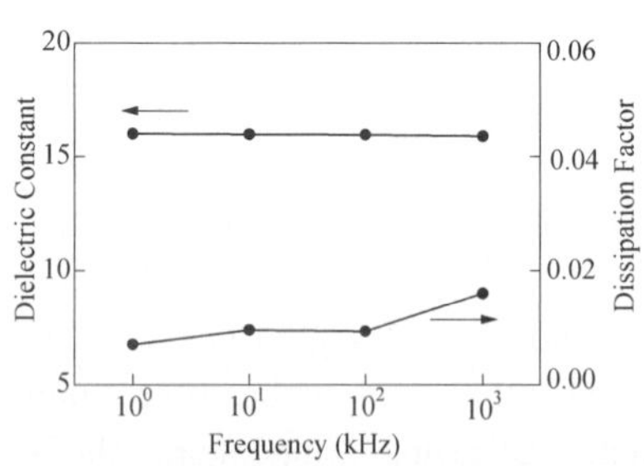

Fig.6 Dielectric constant and dissipation factor of $0.92Ta_2O_5$– $0.08TiO_2$ films annealed at 900°C by RTA as a function of frequency.

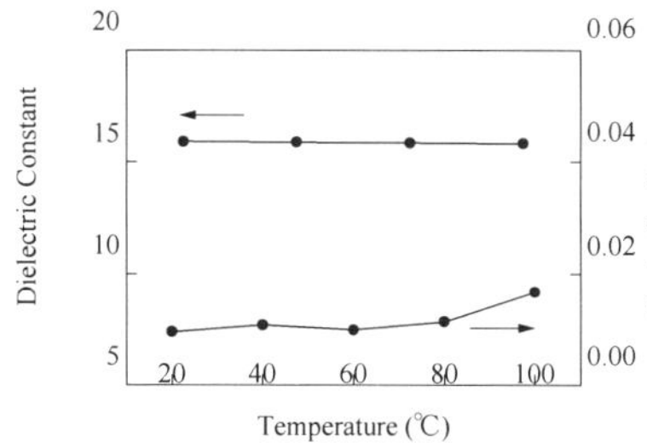

Fig.7 Dielectric constant and dissipation factor of $0.92Ta_2O_5$– $0.08TiO_2$ films annealed at 900°C by RTA as a function of measurement temperature.

frequency. The dielectric constant showed no appreciable dispersion (very little dispersion) with frequency up to 1 MHz indicating that the values of dielectric constant were not masked by any surface layer effects or electrode barrier effects in this frequency range. On the other hand, the loss factor was found to be slight changed with changing frequency. The values of dielectric constant and dissipation factor at a frequency of 1 MHz were about 16 and 0.016 respectively. Figure 7 shows the dielectric constant and dissipation factor of 32-nm-thick $0.92Ta_2O_5$– $0.08TiO_2$ films annealed at 900°C by RTA as a function of measurement temperature. The dielectric constant and the dissipation factor results show, there is no appreciable change with measurement temperature in the range 20-100°C indicating good thermal stability. The change in dielectric constant relative to dielectric constant at 20°C was found to be lower than 1.15% up to 100°C. The dissipation factor was found to increase from 0.008 to 0.014 as the temperature was increased from 20-100°C.

For the suitability of a dielectric material for MOSFET applications, the leakage current characteristics are very important. Figure 8 shows the J/E-$E^{1/2}$ plot for the $0.92Ta_2O_5$–$0.08TiO_2$ thin film. The behavior shown with this plot indicates that the leakage current mainly originates from the Poole-Frenkel conduction. The leakage current density of the film

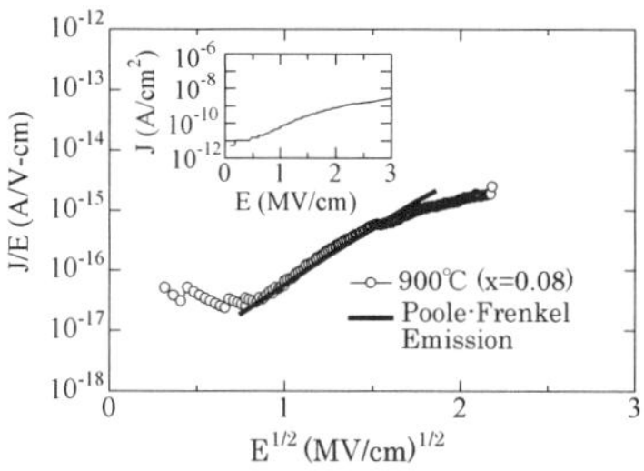

Fig.8 J-E characteristics plotted as J/E versus $E^{1/2}$. Inset shows the J-E curve.

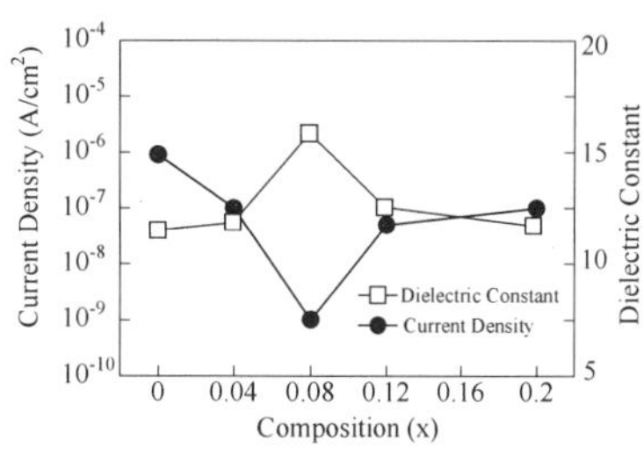

Fig.9 Leakage current density and dielectric constant of $(1\text{-}x)Ta_2O_5$–$xTiO_2$ films as a function of composition.

was as low as 1×10^{-9} A/cm^2 at 3 MV/cm. It is considered that oxygen vacancies are fully compensated by Ti^{4+} ions which are substitutionally incorporated into Ta^{5+} sites in the films [3]. Figure 9 shows the leakage current density and dielectric constant of $(1\text{-}x)Ta_2O_5–xTiO_2$ films as a function of various composition at the temperature of 900°C by RTA in N_2 ambient for 2 min. It is shown that the $0.92Ta_2O_5–0.08TiO_2$ (x=0.08) film is the lowest current density of about 1×10^{-9} A/cm^2 and highest dielectric constant of about 16 which is suggest the suitable for MOSFET applications.

4.Conclusion

Rapid thermal annealing of $(1\text{-}x)Ta_2O_5\text{-}xTiO_2$ thin films with crystalline structure were successfully prepared by metalorganic decomposition. RTA time and ambient do not seem to have a strong effect on structural and electrical properties. The composition of x=0.08 was found to be the best for obtaining the good crystallinity and insulating properties using RTA. A crystalline β-Ta_2O_5 phase was obtained. The surface morphology of the films was smooth and it was found to increase the grain size with increase in RTA temperature and film thickness. The insulating properties of $0.92Ta_2O_5–0.08TiO_2$ thin films were found to be much improved compared to reported Ta_2O_5 thin films [3,5]. The flatband voltage, V_{FB} was range from –0.09 to –0.39V. Interface trap density of 4.5×10^{10} cm^{-2}eV^{-1} was calculated for the sample annealed at 900°C. A charge storage density of 42.5 fC/μm^2 was obtained at an applied electric field of 3 MV/cm. The measured dielectric constant and loss factor at 1 MHz were about 16 and 0.016. The MIS capacitors exhibited good thermal stability. The leakage current density was as low as 1×10^{-9} A/cm^2 up to an applied electric field of 3 MV/cm. The Poole-Frenkel conduction dominates the current-voltage characteristics. The high dielectric constant, low leakage current, low interface trap density, low loss, good thermal stability, and high charge storage density indicate that $0.92Ta_2O_5–0.08TiO_2$ thin films are encouraging for the possibility of dielectric layer in fully scaled ULSI devices.

Acknowledgments

We would like to thank M. Suzuki and K. Tsutsumi of J. A. Woollam Japan Co. Ltd. for VASE measurement. This work was supported by the Foundation of Material Science and Technology (MST Foundation).

References

[1] R. F. Cava, W. F. Peck Jr and J. J. Krajewski, *Nature*, **377**, (1995) 215.
[2] J.-Y. Gan, Y.C. Chang, and T. B. Wu, *Appl. Phys. Lett.*, **72**, (1998) 332.
[3] H. Fujikawa and Y. Taga, *J. Appl. Phys.*, **75**, (1994) 2538.
[4] K. S. Tang, W. S. Lau, and G. S. Samudra, *IEEE Circuits Dev. Mag.*, **13**, (1997) 27.
[5] J. Lin, M. Nakabayashi, A. Tsukune and M. Yamada, *Appl. Phys. Lett.*, **74**, (1999) 2370.
[6] B. H. Lee, L. Kang, R. Nieh, W. J. Qi, and J. C. Lee, *Appl. Phys. Lett.*, **76**, (2000) 1926.
[7] A. Pignolet, G. M. Rao, and S. B. Krupanidhi, *Thin solid films*, **258**, (1995) 230.

[8] S. W. Park and H. B. Im, *Thin Solid Films*, **207**, (1992) 258.

[9] F. C. Chiu, J. J. Wang, J. Y. Lee, and S. C. Wu, *J. appl. phys.*, **81**, (1997) 6911.

[10] P. C. Joshi, S. Stowell, and S. B. Desu, *Appl. Phys. Lett.*, **71**, (1997) 1341

[11] K. Nomura and H. Ogawa, *J. Electrochem. Soc.*, **138**, (1991) 3701.

[12] K. M. A. Salam, H. Konishi, M. Mizuno, H. Fukuda, and S. Nomura, *Short note J. J. Appl. Phys.*, **40**, (2001) 1431.

[13] D. J. Smith and L. Young, *IEEE Trans. Elec. Dev.*, **ED-28**, (1881) 22.

[14] S. C. Sun and T.F. Chen, *IEEE Electron Device Lett,*, **17**, (1996) 355.

[15] Z. G. Zhang, J. S. Liu, Y. N. Wong, Z. S. Zhu, F. Yan, X. B. Chen and H. M. Shen, *Appl. Phys. Lett.*, **73**, (1998) 788.

[16] L. C. Klein, *Sol-Gel Technology for Thin Film, Fibres, Performs, Electronics and Speciality Shapes*, (Noyes, Park Ridge, NJ, 1998).

[17] R. Singh, *J. Appl. Phys.*, **63**, (1988) R59.

[18] Badih El-Kareh, "*Fundamentals of Semiconductor Processing Technology*", IBM Corporation (1995).

[19] J. Y. Sun, C. Wong, Y. Taur, and C. H. Hsu, *VLSI Tech. Digest.*, (1989) P.17.

[20] J. Black, *Proc. Of the 12th Realibility Phys. Symp, IEEE , New York*, (1967) 142.

[21] J. Suarez, B. E. Johnson, and B. El-Kareh, *IEEE Trans. Comp. Manuf. Tech.*, **15**, (1992) 392.

[22] J. S. Mercier, *Solid-State Tech.*, **30**, (1987) 85.

[23] J. R. Hauser, *Proceedings of Material Research Society Symposium*, **525**, (1998) 171.

[24] H. Ono and Ken-ichi Koyanagi, *Appl. Phys. Lett.*, **75**, (1999) 3521.

[25] J. Park and W. Grannemann, *Ferroelectrics*, **10**, (1976) 217.

[26] E. H. Nicollian and J. R. Brews, *MOS (Metal Oxide Semiconductor) Physics and Technology*, (Wiley, New York, 1982).

[27] L. M. Terman, *Solid State Electron.*, **5**, (1962) 285.

Rapid Thermal Processing for Future Semiconductor Devices
H. Fukuda (Editor)

Hard Breakdown Characteristics in a 2.2-nm-thick SiO_2 film

Kenji Komiya, Masahiro Nagahara, and Yasuhisa Omura

High-Technology Research Center and Graduate School of Electronics, Kansai University, 3-3-35, Yamate-cho, Suita 564-8680, Japan
Tel: +81-6-6368-0825, FAX: +81-6-6388-8843, E-mail: omuray@ipcku.kansai-u.ac.jp

1.Introduction

The gate oxide thickness of metal-oxide-semiconductor (MOS) devices is being reduced step by step to match the reductions in integrated circuit scale [1]. The minimum gate oxide thickness is limited by the maximum allowable leakage current and device reliability. It is quite important to understand the degradation mechanisms of ultrathin gate oxide films if we are to design robust MOS devices [2].

For oxide films thicker than 3 nm, it is known that the time-to-dielectric-breakdown depends on electric field applied to the film. Degradation mechanisms have been classified into analog-mode soft breakdown, digital-mode soft breakdown, and hard breakdown (HBD)[3]. Weibull plots are widely used to assess thin-oxide reliability physics since they well integrate cumulative degradation data from many samples. Recent reports suggest that Weibull plots can be made in some cases for oxide films thinner than 2 nm, and that only HBD will be observed in most cases [4].

This paper compares HBD events in constant-current stressed 2.2-nm-thick SiO_2 films formed by various techniques. Background physics of the HBD events are also discussed to interpret phenomena observed.

2.Device fabrication

To examine hard breakdown mechanisms, we fabricated MOS capacitors on n-Si (001) substrates. 2.2-nm-thick oxide films were formed on bare silicon by the rapid thermal oxidation (RTO) technique or the conventional furnace tube oxidation (FO) technique. To create the RTO films, surface oxidation was carried out at 900 C in a dry-oxygen atmosphere for 12 sec. Surface oxidation was carried out at 950 C in a dry-oxygen atmosphere for 23 sec to create 3.3-nm-thick oxide RTO films. 2.2-nm-thick FO films were created by carrying out surface oxidation at 700 C for 20 min for in a dry-oxygen atmosphere. Next, phosphorus-doped poly-Si film was deposited by the low-pressure chemical-vapor deposition technique. The effective gate area was $3x10^4$ μm^2. Since the poly-Si electrode patterns were formed by wet-etching to minimize process-induced damage, we could evaluate the intrinsic properties of ultra-thin oxide films without any extrinsic influence. MOS capacitors have no aluminum electrode.

3.Results and discussion

Gate voltage (V_g) dependencies of initial gate current, stress-induced leakage current (SILC), and post-HBD gate current for several 2.2-nm-thick films are shown in Fig. 1. SILC was observed after constant-current stress (CCS) of 32 A/cm^2 (47 MV/cm) had been applied to the film for 6 hours, while soft breakdown was not observed after additional CCS. Gate voltage was almost constant during CCS. HBD occurred when additional CCS, 100 A/cm^2 (80 MV/cm) for 10 msec, was applied to the film after measuring the SILC characteristics. HBD event was defined by 50% step decrease in gate voltage. The fundamental aspects of gate current characteristics shown in Fig. 1 were reproduced by 100 samples. Regarding these observations, three significant experimental results were noted. (i) Although CCS of 32 A/cm^2 (47 MV/cm) was applied to a 2.2-nm-thick oxide film for 10 hours, HBD was not observed. (ii) When CCS of 40 A/cm^2 (52 MV/cm) was applied to a 2.2-nm-thick oxide film for 10 msec, HBD was always observed. (iii) HBD was not observed when CCS of 32 A/cm^2 (47 MV/cm) was applied to the oxide film for 10 min, followed by CCS of 50 A/cm^2 (60 MV/cm) for 10 min.

Results (i) and (ii) suggest that the HBD of 2.2-nm-thick oxide films is strongly related to electric field stress because CCS conditions for (i) and (ii) are almost identical. So, we investigated the HBD-event probability as a function of electric field stress. The dependence of HBD-event probability on electric field stress is shown in Fig. 2; 10 samples were tested for each stress condition. When the electric field stress is below 50 MV/cm, HBD does not occur, even if CCS under 38 A/cm^2 is imposed for more than 10 hours. This finding is independent of the oxidation technique used.

In the case of 2.2-nm-thick FO films, electric field stress higher than 50 MV/cm always triggers HBD within 10 msec. In the case of 2.2-nm-thick RTO films, the electric field stress must be higher than 57 MV/cm to trigger HBD. For electric field stress ranging from 50

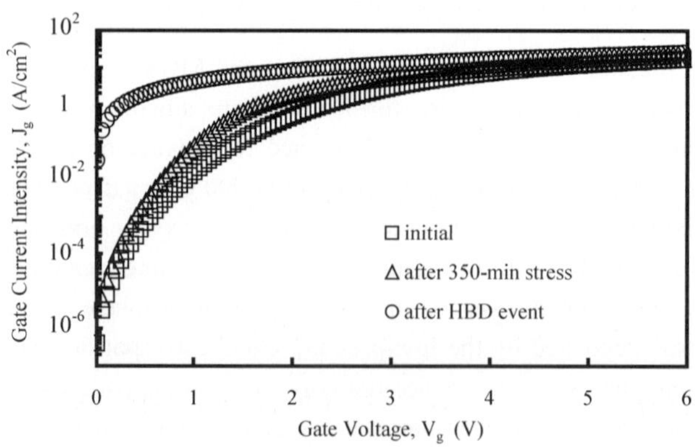

Fig.1 Gate voltage (V_g) dependence of gate current density (J_g) for 2.2-nm-thick SiO_2 film.

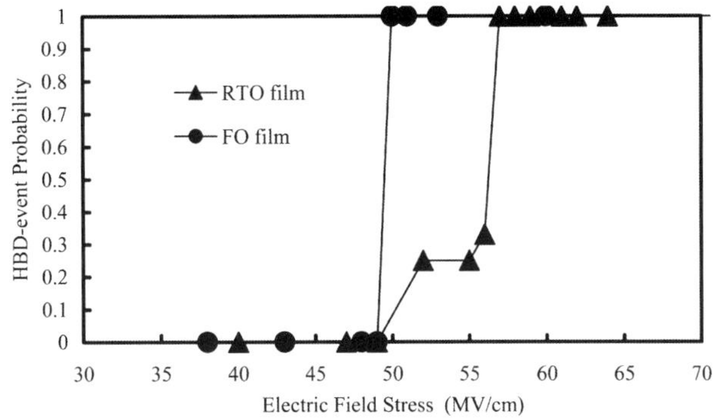

Fig.2 HBD-event probability versus electric field stress

MV/cm to 55 MV/cm, the HBD-event probability has a finite value ranging from 0 to 1; it is considered that this reflects the non-uniformity of oxide quality, including surface roughness. Thus, the critical electric field with regard to HBD is found in Fig. 2.

We must investigate gate voltage behavior under CCS from point of view of film robustness. So, gate voltage dependence on time under the CCS condition of 20 A/cm^2 is shown in Fig. 3 for the RTO film; several devices were tested. It is clearly found that the device does not show the HBD event even after CCS for 3000 sec. So, we have to conclude that 2.2-nm-thick SiO_2 film is inherently robust. Recently, H. S. Momose et al demonstrated the reliability of 2-nm-thick SiO_2 films [5], where electrical stress of about 16 MV/cm was applied to devices and they showed very short lifetime. This should be interpreted that process-induced damage has reduced film lifetime. We also inspected the feature of voltage fluctuation seen in Fig. 3. Fourier transform of voltage fluctuation is shown Fig. 4. The fluctuation power shows typical $1/f^2$ spectra, which strongly suggests that there exist specific defects characterized by a single energy level.

On the other hand, results (ii) and (iii) suggest that breakdown-immunity is improved

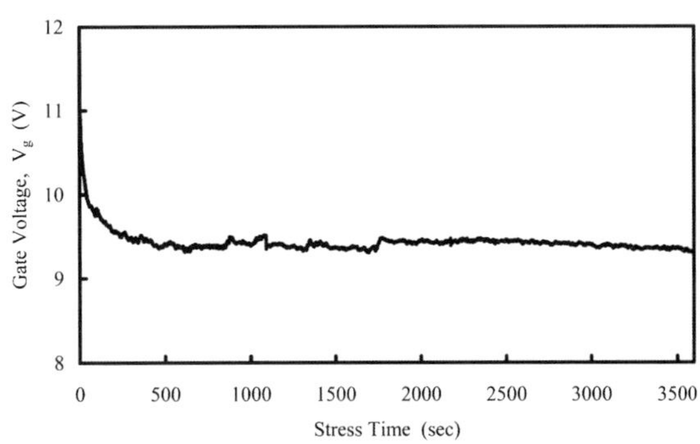

Fig.3 Gate voltage dependence on time under CCS for 2.2-nm-thick RTO film. Constant current for stressing is 20 A/cm^2.

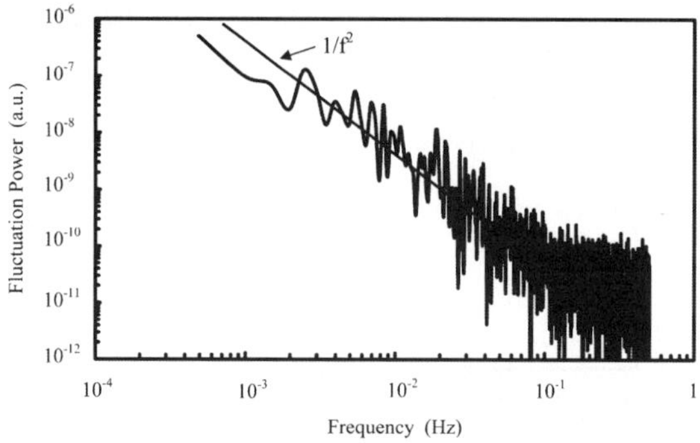

Fig.4 Fluctuation power dependence of gate voltage on frequency under CCS.

when a stress, which is not strong enough to trigger HBD, is applied to the oxide film as a first step. The relationship between stress method and the occurrence of HBD for RTO films is summarized in Fig. 5. In method (A), one-second CCS of 67 A/cm^2 (64 MV/cm) was applied to the oxide film; HBD was always observed.

In method (B), one-second CCS of 67 A/cm^2 (64 MV/cm) was applied to the oxide film after the application of CCS of 36 A/cm^2 (49MV/cm) for 10 min; at this point, however, HBD was not observed. Next, one-second CCS of 100 A/cm^2 (80 MV/cm) was applied to the film; after that HBD was always observed. In method (C), after CCS of 36 A/cm^2 (49 MV/cm) for 600 sec, CCS of 67 A/cm^2 (64 MV/cm) was applied to the oxide film for 5 sec. After that, an optional CCS of 100 A/cm^2 (85 MV/cm) was applied to the oxide film for 40 sec; this method seemed to prevent HBD completely. Similar results were obtained for FO films. Comparing method (C) to method (A) clearly shows that applying a weak initial stress, which does not lead to HBD, improves the films' breakdown-immunity. Comparing method (C) to method (B) indicates that this immunity is improved as the duration period of the weak initial stress increases.

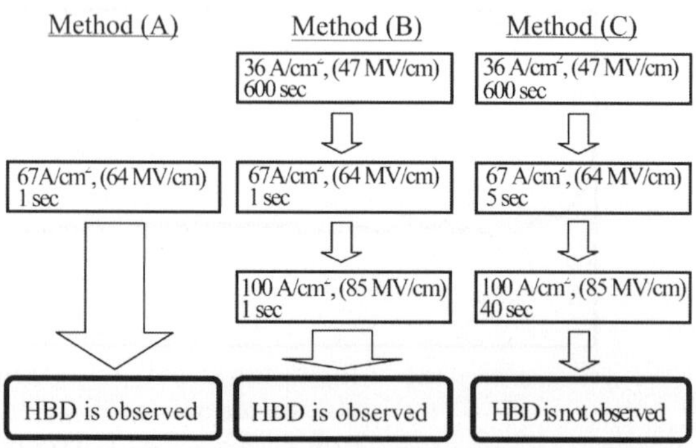

Fig.5 Relationship between stress methods and occurrence of HBD for RTO film

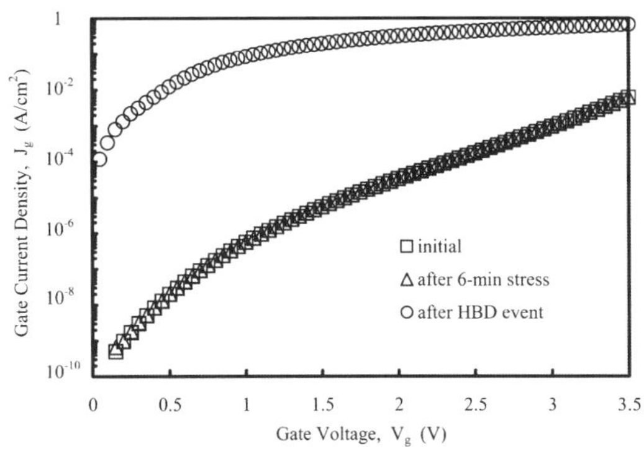

Fig.6 Gate voltage (V_g) dependence of gate current density (J_g) for 3.3-nm-thick SiO_2 film.

So, we discuss the influence of low stress current because weak CCS seldom triggers HBD. When CCS of 38 A/cm^2 (50 MV/cm) is applied to the oxide film, the energy of 660 J/cm^2 is dissipated per second by the stress current. Assuming that the present MOS structure is expressed as a parallel circuit composed of a resistor and a capacitor, the capacitance value is equal to $4.6x10^{-10}$ F, the tunnel resistance value is approximately equal to 1 kΩ and so the charging (or discharging) time constant is estimated to be about $5x10^{-7}$ sec. When we assume that a 5-V step-up is observed at the HBD event, the electrostatic energy of $5.8x10^{-9}$ J is emitted through the capacitor over a period equal to the above time constant ($5x10^{-7}$ sec) [6]; the electrostatic energy of about $4x10^5$ J/cm^2 should be emitted per second. The energy dissipated by the stress current (660 J/cm^2 per sec.) is much smaller than the electrostatic discharge energy. This suggests that the low stress current probably makes traps in the SiO_2 film, but seldom triggers HBD.

In addition, we examined the difference between the breakdown mechanisms of about 2-nm-thick oxide films and over 3-nm-thick oxide films. We consider why the breakdown immunity is improved by applying a weak initial stress to the film. In order to estimate the trap density of thin SiO_2 films, the simple method proposed by D. J. DiMaria et al. is utilized [7]. When we consider the case of a 2.2-nm-thick SiO_2 film (RTO) to which CCS of 38 A/cm^2 (49 MV/cm) has been applied for 6 hours (Fig. 1), the estimated magnitude of created trap density is about 3.7 (arbitrary units). For comparison, gate voltage (V_g) dependencies of initial gate current, stress-induced leakage current (SILC), and post-HBD gate current for 3.3-nm-thick film (RTO) are shown in Fig. 6. SILC was observed after constant-current stress (CCS) of 330 mA/cm^2 (13 MV/cm) had been applied to the film for 360 sec. In this case, the estimated magnitude of created trap density is about 0.058 (arbitrary unit). This is much smaller than the value for the case of a 2.2-nm-thick SiO_2 film, which seems to be a discrepancy. However, this is resolved by the following consideration.

It can be argued that many local traps created in the 2.2 nm film release the stored electrostatic energy fast by trap-assisted tunneling because the effective tunnel resistance of

2.2-nm-thick SiO_2 film is small, which prevents HBD. In the 3.3-nm-thick SiO_2 film, on the other hand, local traps created in the film release the stored electrostatic energy very slowly by trap-assisted tunneling because the effective tunnel resistance of 3.3-nm-thick SiO_2 film is large; relatively high electrostatic energy is stored across the film. Consequently, a local weak spot, in the 3.3-nm-thick SiO_2 film, readily induces the HBD [8].

4.Conclusion

This paper mainly discussed the hard breakdown of 2.2-nm-thick SiO_2 films. It has been shown that the hard breakdown event of a 2.2-nm-thick SiO_2 film greatly depends on the applied electric field. It is strongly suggested that the local weak spots created by applying a low initial stress to a 2.2-nm-thick SiO_2 film resist the onset of hard breakdown. In other words, it is anticipated that the stored electrostatic energy is fast dissipated by trap-assisted tunneling in 2.2-nm-thick SiO_2 film. Consequently, it is strongly suggested that 2.2-nm-thick SiO_2 films are intrinsically quite robust.

References

[1] H. Iwai and H. S. Momose, in 1998 IEEE Int. Electron Devices Meeting Tech. Dig., p. 163.
[2] M. Depas, T. Nigam, and M. M. Heyns, IEEE Trans. Electron Devices, 43(1996)1499.
[3] T. Sakura, H. Utsunomiya, Y. Kamakura, and K. Taniguchi, in 1998 IEEE Int. Electron Devices Meeting Tech. Dig., p. 183.
[4] M. Depas, T. Nigam, and M. Heyns, Solid-St. Electron, 41(1997)725.
[5] H. S. Momose, T. Ohguro, S. Nakamura, Y. Toyoshima, H. Ishiuchi and H. Iwai, Dig. Tech. 2001 Symp. on VLSI Technology (Kyoto, June 2001), p. 77.
[6] S. Ikeda, M. Okihara, H. Uchida, and N. Hirashita, Jpn. J. Appl. Phys., 36(1997)2561.
[7] D. J. DiMaria and J. H. Stathis, Appl. Phys. Lett., 71(1997) 3230.
[8] H. Satake and A. Toriumi, IEEE Trans. Electron Devices, 47(2000)741.

Rapid Thermal Processing for Future Semiconductor Devices
H. Fukuda (Editor)

Rapid Thermal MOCVD Processing for InP-Based Devices

Oleg Kreinin and Gad Bahir

Department of Electrical Engineering and Solid State Institute, Technion-Israel Institute of Technology,
Haifa 32000, Israel
Phone: +972-4-829 3920 Fax:+ 972-4-823 5107 E-mail: ssoleg@tx.technion.ac.il

1.Introduction

The Rapid Thermal Metal Organic Chemical Vapor Deposition technique (RT MOCVD) appears to be very attractive technology in the microelectronics industry for fabrication of thin and high quality semiconductor films with abrupt interfaces [1]. RT MOCVD uses rapid and precise changes in the substrate temperature driven by switching of halogen-tungsten lamps, to control layer growth rather than applying the gas phase switching technique normally used in the standard MOCVD technique. The differences between the two concepts are shown in Fig.1. The main advantages of RT MOCVD system compare with regular MOCVD are as following: -elimination of the pre-processing long temperature wafer exposition step; -the ability to control the growth of abrupt interfaces; and the possibility to realize in-situ several steps of semiconductor device processing.

Selective area epitaxy is an effective tool for the fabrication of optoelectronic integrated circuits. We demonstrated recently a novel mask-less method for the selective epitaxy [2], which uses enhanced deposition of lattice-matched material on a single crystalline surface and

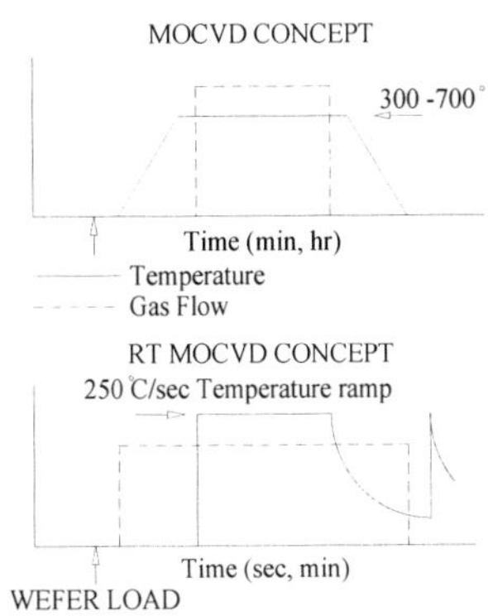

Fig.1 Schematic presentation of the MOCVD and the RT-MOCVD conceptual process schedule.

growth inhibition on ion implanted amorphous surface. One of the expected advantages of this technique is the possibility to anneal the ion implantation damage in-situ and to continue with second growth process.

The goal of this work is to study the possibility of single Rapid Thermal MOCVD system for carry out various process steps for future InP-based laser devices.

2.Basic Growth Technique

InP and InGaAs were grown on Fe-doped, semi-insulating and S-doped (100) InP substrates by the RT-MOCVD technique using A. G. Heatpulse CVD 800 system. This is a low pressure, load lock, horizontal, and laminar flow reactor, heated by two sets of high power halogen lamps (20 lamps of 1.5kW each). Prior to growth, the InP substrates were cleaned using standard process, and immediately loaded into the reactor. TMIn, TMGa, TBA and TBP have been used as materials sources for growth of InP and InGaAs layers. Purified hydrogen was used as the gas carrier.

In the growth of heterostructures as InP/InGaAs it is necessary to prevent mass transport from InP wafer during initial heat-up before growing InGaAs layer, because during this stage InP wafer is exposed to TBA atmosphere. Usually such preservation has been reach either by rapid heat-up or by introducing of small quantities of As_4 [3] along with arsine during the heat-up. Our RT MOCVD system is an ideal facility for rapid heat-up of the wafer but the possibility of using the As_4 in growth ambient of TBA instead of arsine must be proved.

In the preservation experiments the generation of As_4 vapor was produced by a GaAs single crystal wafer, which was placed on the susceptor immediately upstream of the wafer to be preserved.

Figure 2 shows SEM microphotographs of InP wafers after heat treatment in TBA ambient. It is seen that in atmosphere of only TBA (Fig. 2a) InP wafer surface has been attacked by TBA and degraded to some extent. Fig. 2b shows the degradation-free surface, which is a consequence of using As_4 for InP preservation. This surface preservation technique was used for InGaAs layer growth.

a) b)

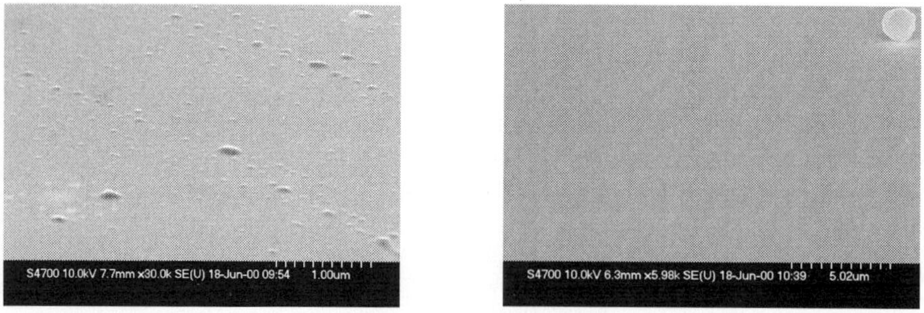

Fig.2 SEM microphotographies of InP wafer surface heat treated in: a) TBA ambient; and b) TBA + As_4 ambient.

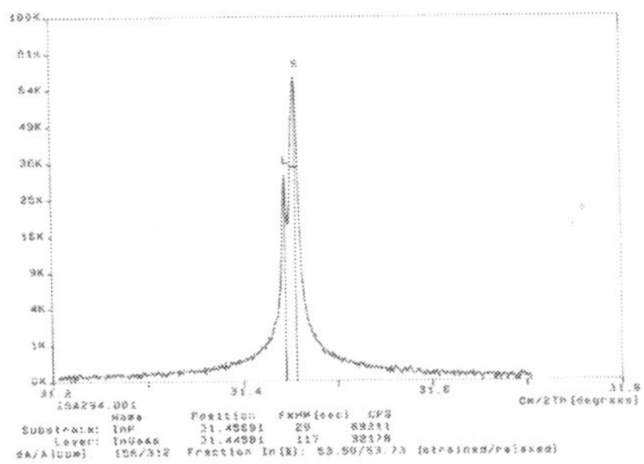

Fig.3 X-Ray diffractometry of undoped InGaAs layer grown on (100) InP.

3.Epitaxial Growth of Undoped InP and InGaAs Layers

At the first step the optimum growth conditions for undoped InP and InGaAs were identified through a set of growth runs, through which the key growth parameters were modified. The temperature was varied in the range 500 - 650°C, the chamber pressure in the range 2 - 25 Torr, growth duration in the range 2 - 20 min and metalorganics flow rates in the range 0.1-100 sccm. High quality InP and InGaAs films (fig. 3) were grown and exhibited an excellent morphology with narrow X-ray full width at half-maximum peak of ~15 arcsec of the InP layer and ~30 arcsec of the InGaAs layer on InP, reflecting a lattice mismatch of 0.02% [4]. The featureless layers were measured to have n type background doping levels of $Nd=5*10^{16}cm^{-3}$ and 300K mobility of 3500 $cm^2/V*s$.

4.RT Growth of Quantum Well Structures

Following the optimization the RT-MOCVD conditions for growth of the single layers of

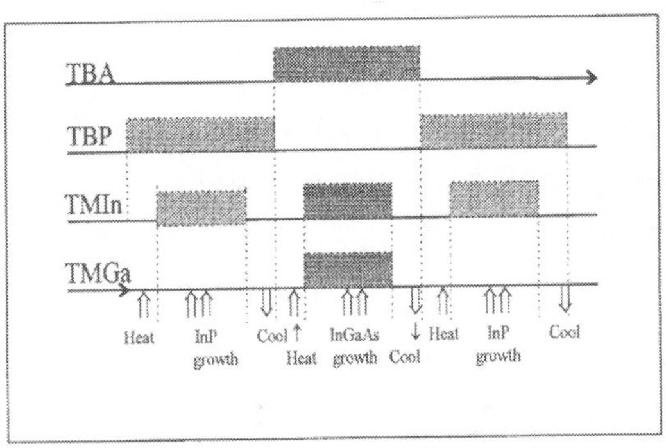

Fig.4 Schematic flow chart of Rapid Thermal -MOCVD growth of InGaAs/InP quantum wells.

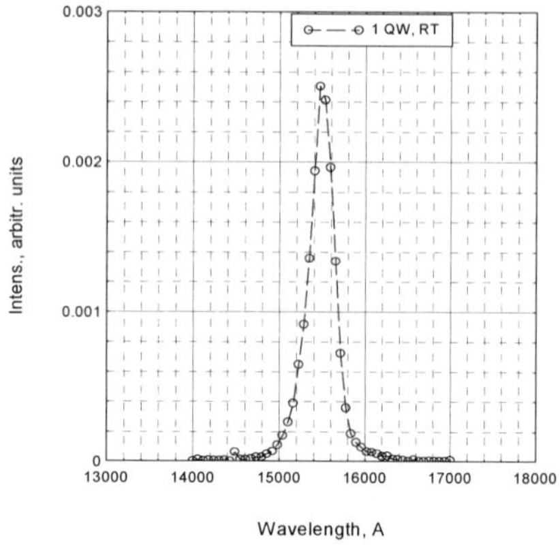

Fig.5 77K photoluminescence spectrum of 1QW InGaAs/InP.

InP and InGaAs, quantum well structures were grown. Figure 4 shows the schematic flow chart for growing InP/InGaAs quantum well structure, deposited on ~0.1 μm thick InP buffer layer grown on exactly (100) oriented semi-insulating InP substrates.

Typical 77K PL spectrum for 100 Å width InGaAs/InP single quantum well is shown in Fig. 5. It is seen that the quality of InGaAs/InP quantum well structures grown by RT-MOCVD system is similar to the widely reported quantum well properties grown by regular MOCVD process.

5.RT Growth of Fe-Doped InP

Semi-insulating (SI) InP layers are of great interest for electronic and opto-electronic devices, particularly as current blocking layers in numerous opto-electronic devices, including lasers detectors and modulator or as a base material for implanted structures.

The RT-MOCVD technique, which has been realized successfully for the growth of undoped InP layers, was used for the epitaxial growth of a Fe-doped InP layers. Beside TMIn and TBP, ferrocene was used as the iron source. The optimum growth conditions in our particular system were found to be at temperature of 650°C, a pressure of 4 Torr, for duration of 15 min at growth rates of up to 2μm/h. The optimum TBP flow rate was 70-80 sccm with TMIn flow rate of 0.15 sccm and ferrocene flow rate of 15 sccm. SIMS depth profile shows uniform Fe incorporation in InP for samples grown in optimum conditions with atomic concentration of about 10^{18} cm^{-3}. The specific resistivity of the grown semi-insulating InP layer was measured to be $\rho > 5x10^7$ Ωcm (fig. 6), about the same order of magnitude as for the best commercial SI InP substrate. Details of growth conditions and electrical characterization technique can be found in [2].

6.Selective RT-MOCVD Growth

Selective growth of InP and related materials on patterned planar and non-planar substrates is one of the key steps in the fabrication of discrete and integrated optoelectronic devices. In this work we continued the study of ion implanted mask-less technique for selective growth [2]. This technique makes use of the fact, that the growth of InP on the amorphous implanted area is inhibited due to the lack of lattice constant definition. We combined the above technique to form buried layers by the following processing sequence steps: a) Ion implantation for mask-less selective area growth definition. b) First selective growth of InP layer on the non-implanted features. c) Growth of the second selective InGaAs layer on the top of the first InP layer. d) *In- situ* Rapid Thermal annealing of ion implantation damage. e) Fe doped InP re-growth over the whole area. Thus, the first, selective growth is completely covered by the burying re-growth layer.

At the first step the optimum conditions for ion-implanted amorphization of InP surface were investigated. Fe ion looks promising for both: as an effective heavy ion for modification of the unmasked surface areas, and as an ion for semi-insulating doping for further semiconductor structure fabrication. The Fe ion implantation was performed at room temperature with doses between $1x10^{14}$ to $4x10^{14}$ ions/cm^2 at energy of 100 and 200 keV and a constant current density of 0.1μA/cm^2. Implant damage in InP can consist of either amorphous layers or extended crystalline defects such as dislocations and stacking faults, depending on the ion, the dose, the dose rate and the implant temperature [4,5]. For the doses we used in this work the InP surface reached the amorphization conditions [5].

In our experiments we used implantation masks as a parallel stripes with filling factor (the percentage of masked area to all area of the surface) ~50%. Fig. 7 shows microphotography of the selective InP growth on Fe- implanted InP substrate. It is seen, that there is no growth on

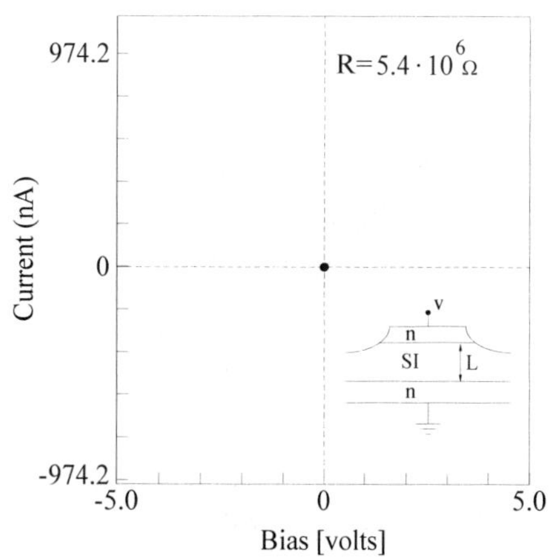

Fig.6 Measured I-V curves for the Fe-doped InP layers.

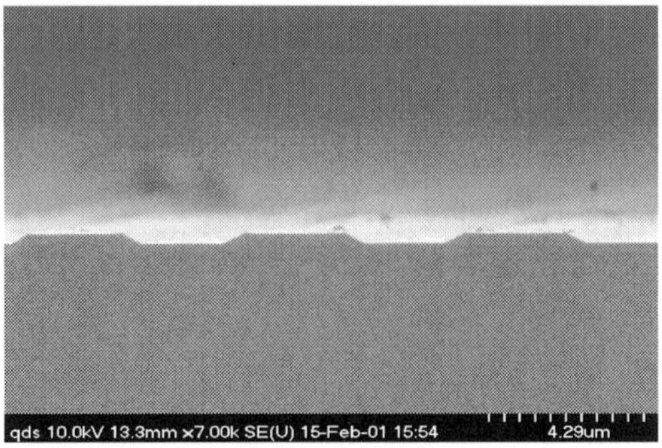

Fig.7 SEM microphotography of the selectively grown InP layer on Fe-ion implanted InP substrate.

the implanted areas. The variation of the ion implantation dose showed that selectivity is achieved in full-investigated range but it is better to use the minimum dose needed to amorphization to facilitate future Rapid Thermal Annealing [2].

We studied a range of selective growth runs through which the main selective growth parameters were varied. The optimum values of temperature and pressure for selective growth of InP and InGaAs layers were about the same as for regular RT MOCVD growth (see above). The key parameters for receiving better selectivity are TMIn and TMGa flows and growth rate. According to our experience, the low metalorganics flow rate leads to a better selectivity. If for regular growth we were using 60-75 sccm of TMIn, for selective growth the best results were achieved at 30-35 sccm of TMIn; instead of using 10-15 sccm of TMGa for selective growth ~5 sccm of TMGa is optimal. The growth rate for selective InGaAs growth must be 2 times more compare to regular RT MOCVD growth. Fig. 8 shows the microphotography of the second selective InGaAs layer grown on InP selective layer.

The optimum values of temperature and pressure for selective growth of InGaAs were about the same as for regular RT MOCVD growth of InGaAs as described above.

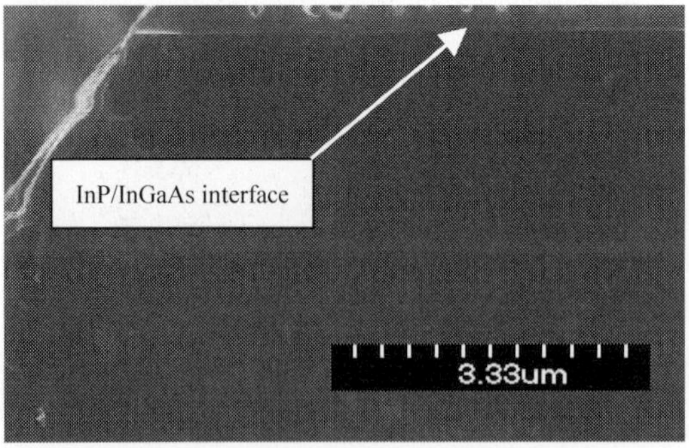

Fig.8 Microphotography of selectively grown InP/InGaAs heterostructure.

7.RT-Annealing

The re-growth of semi-insulating InP:Fe around mesas for current confinement has to be rapid enough to avoid excessive quantum well mixing and unwanted dopant diffusion. An attractive option of the RT-MOCVD technique is to bury the first selective growth structure by a second non-selective processing step. For achieving this step the implanted amorphous region layer should be annealed and re-crystallize. For reducing of ion-implanted damage in-situ Rapid Thermal Annealing (RTA) technique was used. It is known that recrystalization and implant activation in implanted InP and related materials requires temperature above the incongruent evaporation point. This means that some kind of surface protection must be employed during the RTA process [5,6]. This protection supplies in this work by the TBP overpressure in the RT-MOCVD chamber.

The optimum RTA conditions in our RT MOCVD system were achieved at temperature of ~750°C, for 15 seconds with temperature ramp of 200°C/s, pressure of ~ 80 Torr in atmosphere of 200 sccm of TBP. In such conditions we reached complete recrystalization and implant activation without wafer surface damages [4].

8.Regrowth of a burying semi-insulating InP layer

The last stage of this study is an effort to combine the three steps that described above: *i*) ion implanted mask-less selective growth of InP/InGaAs heterostructure *ii*) *in-situ* rapid thermal annealing of ion implantation damage and *iii*) *in-situ* re-growth of burying Fe doped semi-insulating layer all over the sample. Following ion implantation and sample preparation, the whole process was done in three successive steps in one growth process at the same conditions that described in the above paragraphs.

Fig. 9 shows the microphotography of the selectively grown InGaAs layer and the final regrown Fe-doped burying InP layer following the rapid Thermal Annealing.

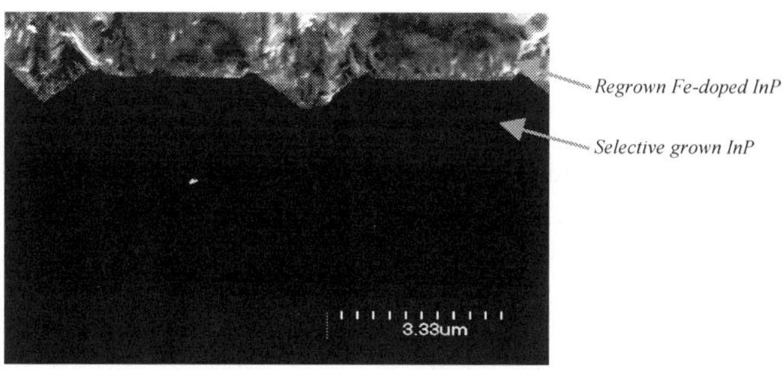

Fig.9 Microphotography of the regrown Fe-doped burying InP layer following RT Annealing.

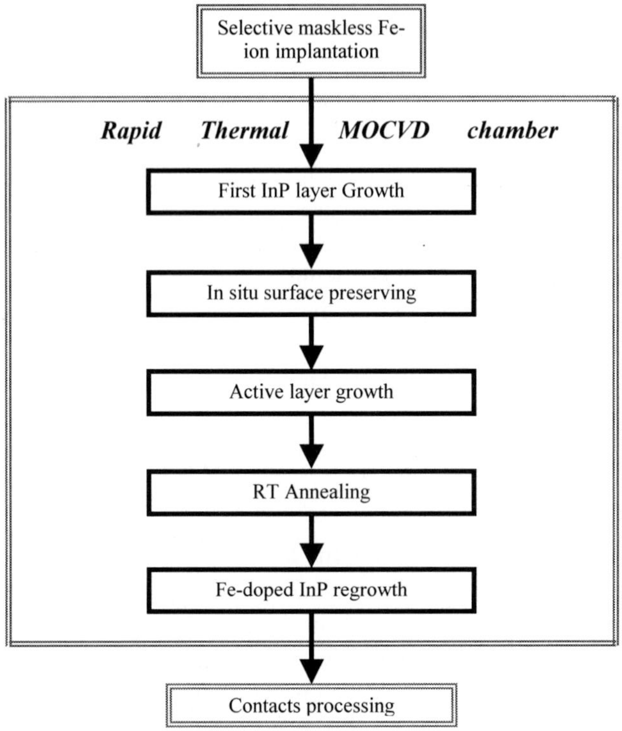

Fig.10 Single chamber technique for future InP-based device fabrication.

4.Conclusions

We have demonstrated the potential of RT MOCVD technique for carry out the various processes associated with the integrated processing of InP-based devices. As an conclusion we would like to present the following schematic (fig. 10) describing future InP- based device processing by means of single chamber RT MOCVD system.

Acknowledgement

The research was supported by "Center for Absorbtion in Science", Ministry of Immigrant Absorbtion, State of Israel.

References

[1] A. Feingold and A. Katz, *Mateials Science and Engin.*, **R13**, No. 2, (1994) 57.

[2] O. Kreinin and G. Bahir, *IPPRM 1999 Proc.*, (1999) 59.

[3] R. Bhat, et al, *J. Crystal Growth*, **107**, (1991) 871.

[4] O. Kreinin and G. Bahir, *IPPRM 2001 Proc.*, (2001) 401.

[5] G. Bahir, J.L. Merz, J.R. Abelson and T.W. Sigmon , *J.Appl. Phys.*, **65**, (1989) 149.

[6] S. J. Peatron and U. K. Chakrabarti, *Indium Phosphide and Related materials*, Artech House, (1992) 211.

Rapid Thermal Processing for Future Semiconductor Devices
H. Fukuda (Editor)

Sb Pile-up at the SiO_2/Si Interface during Drive-in Process after Predeposition using SOG Source

Tomohiro Ichino, Hideo Uchida, Masaya Ichimura and Eisuke Arai
Department of Electrical and Computer Engineering, Nagoya Institute of Technology, Gokiso-cho, Showa-ku, Nagoya 466-8555, Japan
Phone:+81-52-735-5581 Fax:+81-52-735-5442
Email:ichino@hermite.elcom.nitech.ac.jp

1.Introduction

Sb is promising as dopant for shallow junction formation in scaled down devices because of its low diffusivity. Recently, 0.1 μm MOSFETs with ultra-shallow source and drain extensions fabricated by Sb implantation have been reported [1]. In another report, it has been described that pile-up of Sb at the SiO_2/Si interface results in dopant loss during the solid phase epitaxial growth of amorphized layer formed by the implantation[2]. However, fundamental data of Sb diffusion are shorter than other dopants. In addition, many annealing steps in the LSI fabrication process and reactions at Si surface have great influences on shallow junction formation. Therefore it is necessary to collect and analyze fundamental data of Sb diffusion for accurate diffusion modeling.

The purpose of the present study is to clarify the Sb pile-up phenomenon during a drive-in process after predeposition using a diffusion source and to measure the diffusion coefficient of Sb in Si.

2.Experimental Procedure

Substrates used were p-type (100) Cz Si wafers with resistivity of 0.1-1 Ω cm. The Sb predeposition was carried out at 1000°C for 30 min in N_2, using SOG (spin-on-glass composed of SiO_2 and Sb_2O_3) source which was baked at 120 °C for 1 hr in air. After removal of the source by HF, the drive-in diffusion was carried out at 1000°C for various times in N_2, O_2, and wet O_2 atomospheres. During the drive-in diffusion, oxide layers were formed not only in O_2 and wet O_2 but also in N_2, as described below. Sample preparation and measurements are illustrated in Fig. 1. Sb depth profiles in both Si and oxide were obtained by the secondary-ion mass spectroscopy (SIMS: CAMECA IMS-4f) analysis. The primary beam was Cs^+ and its energy and current were 10 keV and 20 nA, respectively. The sputtered depths were obtained by measuring the crater depth with a stylus profilometer assuming that the sputtering rate remained constant. The Sb profiles in Si were measured after the oxide layers were removed by HF dipping. For measurement of profiles in oxide, the oxide was

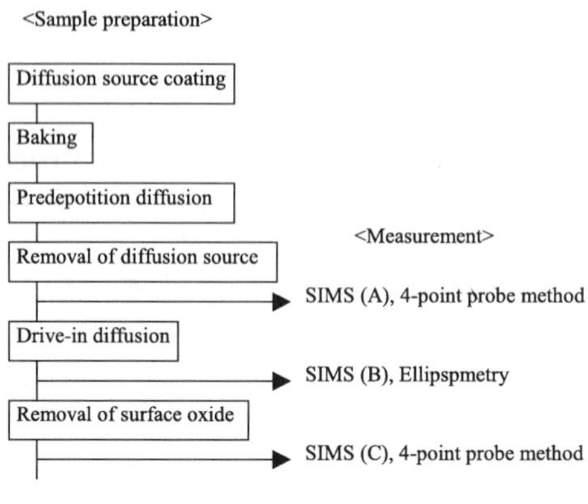

Fig.1 Sample preparation and measurement

etched down to the thickness of about 0.1 μm to prevent charge-up effect. Sheet resistance of diffused Si was measured by the 4-point probe method. Thickness and refractive index of the oxide layers formed at the Si surface during the drive-in diffusion were measured by ellipsometry.

3.Results and Discussion

3.1 Pile-up of Sb at the interface

Figue 2 shows Sb depth profiles in Si for predeposition for 30 min and drive-in diffusions for 30 min in N_2, O_2, and wet O_2 (SIMS (A) and (C) in 1, respectively). After the drive-in diffusion, the surface concentration decreases dramatically in all atmosphere. In addition, the depths for O_2 and wet O_2 drive-in diffusions were shallower than the depth for predeposion diffusion. Table I shows sheet resistance and Sb total amounts in Si obtained from the depth profiles in Fig. 1. The total amount of Sb in Si after the drive-in diffusion decreases with increase in the oxide thickness shown in Table II Compared with predeposition, the Sb

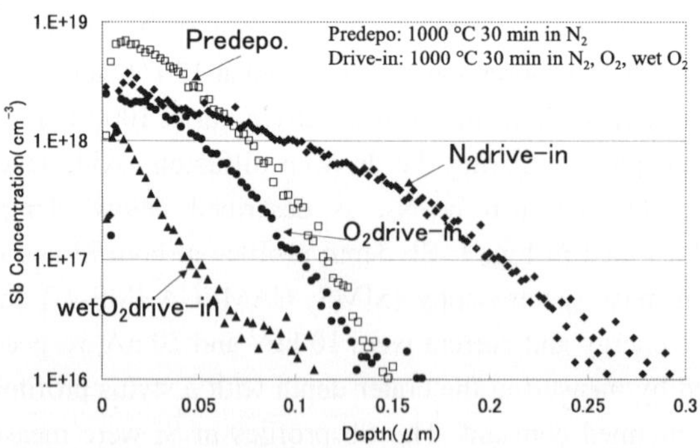

Fig.2 Sb depth profile in i (SIMS (A), (C) in Fig. 1).

Table I. Sheet resistance and total amounts of Sb in Si obtained from the depth profiles of Fig. 1

	Sheet resistance (Ω/□)	Sb amounts in Si (cm^{-2})
Predeposition	1730	$3.12x10^{13}$
$_2$ drive-in	1880	$1.65x10^{13}$
O_2 drive-in	1990	$1.17x10^{13}$
wet O_2 drive-in	2500	$3.03x10^{12}$

Table II. Thickness and refractive index of the oxide layer on Si surface after drive-in diffusion in N_2, O_2 and wet O_2

	Oxide thickness (μm)	Refractive index
$_2$ drive-in	0.0097	1.47-i0.12
O_2 drive-in	0.065	1.47
wet O_2 drive-in	0.27	1.47

amount loss in Si is 27% and 64% and 91% for N_2, O_2 and wet O_2 drive-in diffusion, respectively. This result corresponds well to increase in sheet resisitance.

Table II shows thickness and refractive index of the oxide layers on the Si surfaces after the drive-in diffusion for 30 min in N_2, O_2 and wet O_2. Oxide thickness for the N_2 drive-in diffusion was 9.7 nm. The reason for the oxidation in N_2 drive-in is explained as follows; the Si surface is slightly oxidized during loading/unloading of the wafer into the diffusion furnace because of O_2 pulling-in from the air. Only for the N_2 drive-in diffusion, refractive index has an imaginary part, which suggests the presence of elemental Sb in the oxide layer.

In order to investigate the cause of decrease in the total amount in Si and increase in sheet resistance, we obtained Sb depth profiles in oxide layers by SIMS.

Figure 3 shows Sb depth profiles in the oxide (SIMS (B) in Fig. 1) combined with those in Si (SIMS (C) in Fig.1). For the wet O_2 drive-in, the oxide thickness is reduced to about 0.1 μm by etching. The position of the interface between oxide and Si (shown by the dashed line) is determined according to the oxide thickness shown in Table II. When we estimate the value of Sb concentration in the oxide, we cannot assume that the secondary ion yield of Sb in the oxide is the same as that in Si because of the matrix effect. So, to obtain Sb depth profiles in the oxide, ion intensity of Sb was normalized by that of Si and was converted to the Sb concentration assuming that Sb did not out-diffuse into the ambient during the drive-in diffusion. This assumption may be valid, because the oxide grows at the Si surface so that out-diffusion of Sb is prevented. From Fig. 3, Sb pile-up is clearly observed at the oxide side

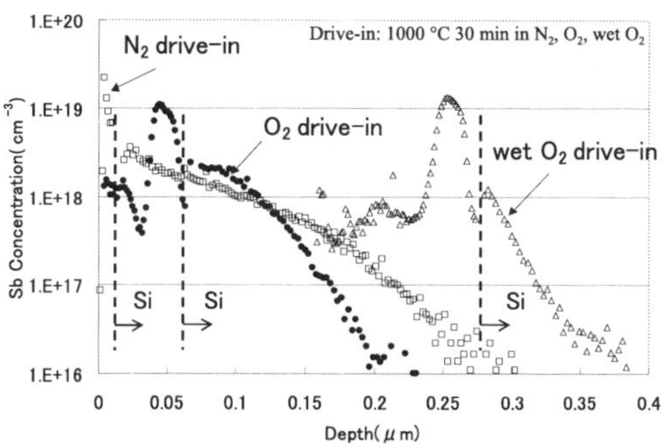

Fig.3 Sb depth profiles combining those in the oxide (SIMS (B) in Fig. 1) and those in Si (SIMS (C) in Fig. 1).

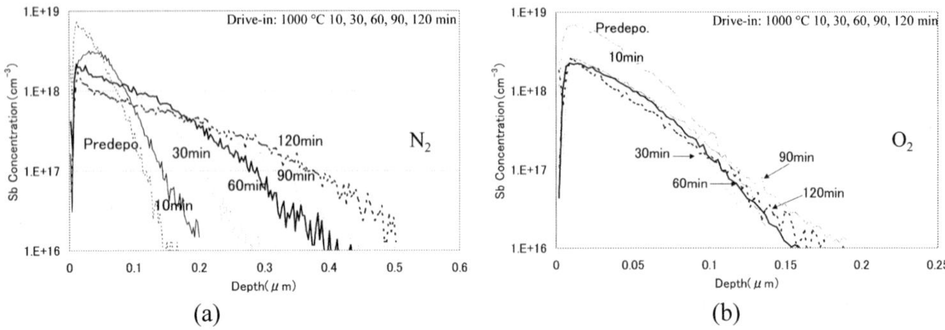

Fig 4 Time dependences of the Sb drive-in diffusion (SIMS (C) in Fig. 1). Diffusion atmospheres are (a) N_2 and (b) O_2.

of oxide/Si interface in all the ambient. From Figs. 2 and 3, it is found that the Sb pile-up disappears after dipping in HF. Therefore it is expected that Sb piles up in a form soluble in HF and not in Si. This pile-up in oxide causes large decrease in Sb amount in Si and increase in sheet resistance during the drive-in diffusion. In addition, Sb peak concentration of the pile-up layer is of the order of 10^{19} cm^{-3}. Since the Sb surface concentration in Si is of the order of 10^{18} cm^{-3}, Sb concentration of the pile-up layer is much larger than surface concentration in Si.

The present results are striking because it is well known that the segregation coefficient of Sb reported previously is much larger than unity (concentration in Si / concentration in SiO_2 > 1) [3], i.e., the SiO_2 layer is expected to reject Sb. Furthermore, recently, It has been reported that the pile-up is observed only for samples with implanted Sb concentrations above 10^{20} cm^{-3}, larger than the solid solubility limit [2]. However, in this preset study, Sb is doped into Si in the predeposition process but not ion-implantation, and thus its concentration never exceeds the solid solubility. From these experimental results, it is confirmed that during the drive-in diffusion, Sb is extracted from Si and piles up in the oxide side of the interface, irrespective of doping method and Sb concentration.

3.2 Sb diffusion in Si

Figure 4 shows Sb drive-in diffusion profiles in N_2 and O_2 (SIMS (C) in Fig. 1). The drive-in diffusion was carried out at 1000 °C for 10, 30, 60, 90 and 120 min. For the N_2 drive-in shown in Fig. 4(a), the oxide thickness formed during the drive-in diffusion was about 100 Å, which was independent of diffusion time. The Sb depth profiles in Si show Gaussian shape except for the near-surface region. While, the profiles for the O_2 drive-in in Fig. 4(b) show little expansion. It is well known that Sb diffuses with vacancy mechanism [4] and therefore, Sb diffusivity is suppressed by injection of interstitial Si due to surface oxidation. The shift of the oxide/Si interface with surface oxidation also results in decrease in the apparent diffusion depth. Figure 5 shows diffusion time dependences of sheet resistance and Sb amount in Si obtained from the depth profiles shown in Fig. 4. The Sb amount in Si decreases during initial

30 min and then remains almost constant in both atmospheres. This fact shows that the Sb pile-up layer is formed during initial 30 min.

In the present diffusion condition, the Sb diffusion is performed under the intrinsic condition (i.e., the Sb concentration is lower than the intrinsic carrier concentration at 1000 °C). Therefore, the diffusivity of Sb is expected to be constant in the concentration region. The Sb amount in Si remains almost constant beyond 30 min in both atmospheres, as described above. Thus, we can assume that the Sb diffusion profile beyond 30 min shown in Fig. 4 is expressed by the Gaussian distribution. Generally, the Gaussian distribution is expressed by the following equations,

$$C(x,t) = \frac{Q}{\sqrt{\pi Dt}} \exp\left(-\frac{x^2}{4Dt}\right) \quad (1)$$

$$Dt = D_{pre}t_{pre} + D_{dri}t_{dri} \quad (2)$$

where Q is the dopant amount in Si, D_{pre} and D_{dri} are the diffusion coefficients for the predeposition and drive-in diffusion, t_{pre} and t_{dri} are the time of the predeposition and drive-in diffusion, respectively. Here D_{pre} is estimated to be 6.0 x 10^{-15} cm^2/s by fitting the measured profile in Fig.1 to the erfc distribution. Figure 6 shows the comparison of the experimental (symbols) and simulated (lines) profiles for the N_2 drive-in diffusion. From this figure, the Gaussian profiles can be fitted well to the experimental profiles beyond 30min. In this fitting, D_{dri} is estimated to be 1.85 x 10^{-14} cm^2/s. Sb amount in Si continues to decrease within 30min due to the pile-up formation as shown in Fig. 5. Thus, we can not assume that the depth profiles at 10 and 30 min are expressed by the Gaussian distribution. If the depth profiles at

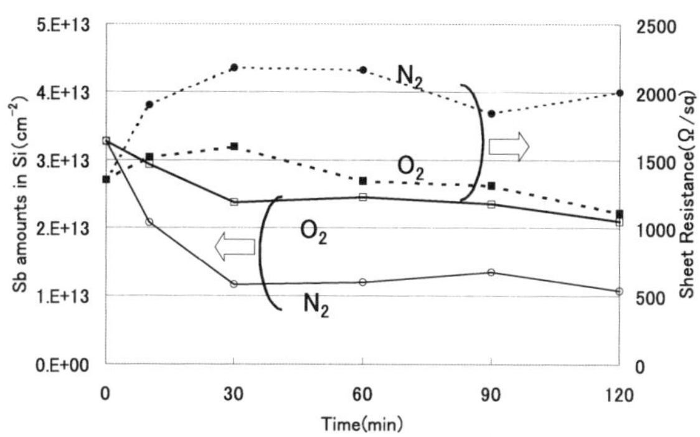

Fig.5 Diffusion time dependences of sheet resistance (dashed lines) and Sb amount in Si obtained from the depth profiles of Fig. 4(solid lines).

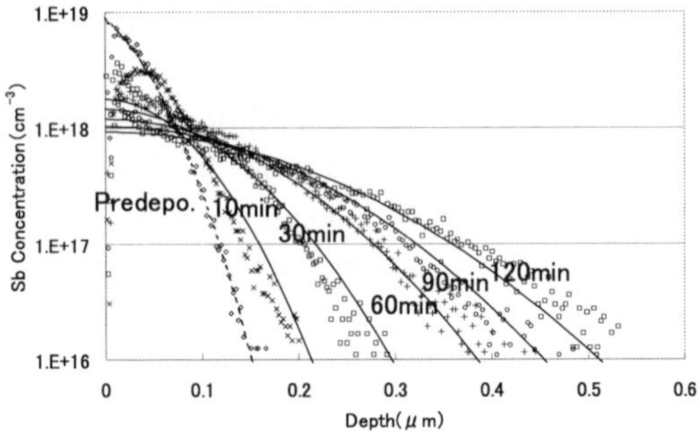

Fig.6 Comparison of the experimental (symbols) and simulated (lines) profiles for the N_2 drive-in and predepotision diffusion.

10 and 30 min is simulated by Gaussian distribution, the diffusion coefficient will become much smaller than that beyond 30 min.

From these experimental results, we can simulate the diffusion profiles of the N_2 drive-in diffusion beyond 30 min at 1000°C using Gaussian distribution. The diffusion coefficient beyond 30 min shows several times larger than that of predeposition. It is possibly considered that the different diffusion coefficients are caused by decrease in the vacancy concentration in Si by injection of the interstitial Si during the oxidation, i.e., Sb diffusion is suppressed by injection of interstitial Si. For the predeposition process, the elemental Sb is generated at the Si surface by the following interface reaction and at the same time, the Si surface is oxidized.

$$1/2Sb_2O_3 + Si \rightarrow Sb + SiO_2 \qquad (3)$$

For the N_2 drive-in process, about 100 Å-thick oxide is also formed during initial period as described above. These surface oxidation processes are considered to be one of the origins of the decrease in diffusion coefficient.

4.Conclusion

We investigated the Sb pile-up phenomenon and the diffusion coefficient during the N_2 drive-in diffusion. Sb piles up rapidly within 30 min at oxide side of the oxide/Si interface during drive-in diffusion in N_2, O_2 or wet O_2 irrespective of doping methods, i.e., implantation or thermal diffusion. The Sb pile-up is generated even for concentration below solid solubility of Sb. The pile-up is a major cause of decrease in diffused amount of Sb in Si and increase in sheet resistances. The pile-up amount of Sb is dependent upon diffusion ambient.

The diffusion coefficient of the N_2 drive-in diffusion increases rapidly within 30 min and then saturates. The diffusion coefficient beyond 30 min shows several times larger than that of predeposition.

References

[1] K.Shibahara, M.Mifuji, T.Kugimiya, H.Tsuno, S.Yokoyama, M.Nagata, S.Miyazaki and M.Hirose, Tech. Dig. IEDM, p.579 (1996)

[2] K.Shibahara, H.Furumoto, K.Egusa, M.Koh and S.Yokoyama, Proc. MRS. Symp. Vol.532, p.23 (1998)

[3] A.S.Grove, O.leistiko and C.T.Sah, J. Appl. Phys., Vol.35, p.2695 (1964)

[4] S.Mizuo and H.Higuchi, J. Electrochem. Soc., Vol.130, p1942 (1983)

Rapid Thermal Processing for Future Semiconductor Devices
H. Fukuda (Editor)

Large Refractive Index C-S-Au Composite Film Formation by Plasma Processes

Masaki Matsushita, Md. Abul Kashem and Shinzo Morita

Department of Electronics, Graduate School of Engineering, Nagoya University, Nagoya, Japan
Phone:+81-52-789-5722 Fax:+81-52-789-5723 E-mail: matussita@tcp-ip.or.jp

1.Introduction

Optical interconnection between large-scale integrated circuits is important to minimize the time lag of signal transportation in the interconnection circuits. Recently Sasaki et. al. reported practical free-space optical interconnection circuit which is 40 mm long, 10 mm width and 3 mm thickness [1].

For the optical interconnection circuit, Wada and Kimering discussed a new approach to use a high-index contrast optical wave-guide and a photonic crystal wave-guide in order to minimize the optical system [2]. The high-index contrast optical wave-guide is a conventional optical wave-guide, but the refractive index of wave-guide must be large as possible as it can be formed in order to minimize the curvature. The material also must be transparent at the wavelength. With using Si core and air clad, they realized a curvature radius of 3μm at 0.5 dB level. However the light propagation loss will increase with increasing the refractive index and decreasing a curvature radius. It is supposed that the phtonic crystal overcomes the difficulty. The photonic crystal wave-guide can be realized according to the photonic band-gap (PBG) in the periodicity of refractive index material [3,4].

Large refractive index and optically transparent materials are necessary for a photonic crystal fabrication. Widely using materials are GaAs, SiO_2 and related materials. However some of them are not suitable for plasma etching process and some of them are not high refractive index material. Presently fabricated photonic crystals are usually using air clad [5] because the high refractive index materials have not good process compatibility and also there are few materials to fulfill high refractive index contrast. Therefore we proposed C-S-Au composite film as a new large refractive index material [6, 7] and amorphous hydrocarbon film as a small refractive index material, however the most of composed atoms become volatile under oxygen plasma treatment. Only Au is chemically inert, but Au is also known to be physically etched slowly in the plasma [8]. Whereas it will be concluded that the proposed materials have process compatibility.

In this paper, C-S-Au composite film formation will be reported and the optical properties and processes for 2-dimentional photonic crystal (2-D PC) fabrication will be also discussed.

2.Experimental

Co-operation process of plasma CVD and sputtering was used for the C-S-Au composite film formation with using a parallel discharge electrode system at 13.56 MHz. The upper electrode is made of graphite to eliminate contamination from the electrode because the upper electrode is negatively biased under the RF plasma condition and the electrode material is sputtered during the plasma CVD. C-S compound will be formed with using CH_4, SF_6 and Ar mixture gas. When Au plate with area of 5×5 cm^2 is set on the upper discharge electrode, Au atom is mixed into the film on the lower electrode by sputtering from the upper electrode. As substrates, Corning 7059 glass and Si wafer were used for the characterization of the film.

The films were characterized by measuring thickness, ESCA, optical transmittance, and refractive index at a wavelength of 632 nm. The etching rate was also measured as a function of the discharge power by using RIE apparatus at 0.1 Torr, an O_2 gas flow rate of 10 SCCM and a frequency of 13.56 MHz.

3.Results and Discussions

C-S and C-S-Au composite films were prepared at a fixed condition of gas pressure of 0.1 Torr, discharge power of 100 watt and discharge duration of 30 min, where gas flow rates of CH_4, SF_6 and Ar were controlled independently. Initially the films were formed at a fixed flow rate of 10 SCCM for Ar, 10 SCCM for CH_4 but SF_6 flow rate was changed from 2 to 25 SCCM. The atomic compositions of the films evaluated by ESCA measurements were shown in Table 1. The atomic compositions of hydrogen and fluorine were neglected because they were very small compared to other atoms. However Au atom content was listed in Table 1 even if the content is smaller than 1 atomic % because the contribution of Au on the refractive index is significantly large compared to hydrogen and fluorine atom [7].

From Table 1, we concluded that CH_4 and SF_6 were dissociated to C and S atom by forming HF during the plasma CVD process and the hydrogen and fluorine atom were eliminated from the reactor in the form of HF by the pump. However there are large deviation from the stoichiometry of reaction in the gas composition of CH_4 and SF_6. The deviation is supposed that intermediate products like as C_2H_x and SF_4 are also stable in the plasma and evacuated from the reactor.

Refractive indexes of C-S and C-S-Au film at about 0.6μm are plotted in Fig. 1. 2.4 is the maximum value for C-S film and 3.7 is the maximum value for C-S-Au film. Compared to diamond refractive index of 2.38 [9], the values are larger. Refractive index of GaAs is know to be 3.4 at an optical wavelength of 1 μm [4].
Therefore the value of 3.7 is not small as a photonic crystal material. The polymer and plasma polymerized film showed 1.5-1.6 as refractive index usually [10]. Therefore the index contrast between the C-S-Au and carbonaceous (aC) film is about 2.1, which is profitable value for the photonic crystals.

Table 1 Atomic composition of C-S and C-S-Au films

Flow Rate of SF_6	Atomic Composition (%)					
	C-S film			C-S-Au film		
	C	S	Au	C	S	Au
2	98.8	1.12	0	96.0	1.86	2.1
5	93.1	6.90	0	96.0	3.22	0.7
10	93.2	6.80	0	93.4	5.84	0.7
15	90.6	9.40	0	88.3	11.6	0.0
20	85.2	14.8	0	85.9	14.0	0.1
25	79.3	20.7	0	81.0	18.7	0.1

Co-operation processes of plasma CVD and sputtering were studied by Biederman, Martinu, Kay and other [10]. They observed the metal was contained in the film in the form of clusters. Martinu measured the refractive index to be 2.7 at 0.7μm for fluorinated polymer with Au clusters. The refractive index was calculated theoretically, where the metal clusters were expected to cover by dielectric layer.

The large refractive index of our film cannot be explained by Martinu's model on metal cluster covered by polymer. Electronic polarizabilities of atoms in the film were evaluated with using the refractive indexes measured and the atomic composition in the film [7]. The large refractive index of C-S-Au composite film was referred to the atoms with the large atom number, namely S and Au atom. Then our C-S-Au composite film was expected that Au atom was not clustered and was distributed atomically in the film. However there is a question why Au atom was uniformly distributed in our film instead of cluster formation. One possibility isthat the atomic density is very small compared to that Martinu reported. Another possibility is referred to chemical bond between S and Au atom because S and Au can be bonded weekly even if Au atom is inert material [11, 12].

Next problem is optical transmittance of C-S-Au film. Optical transmittance of aC, C-S and C-S-Au film are shown in Fig. 2. The aC is almost transparent in the measured wavelength range. However the C-S and C-S-Au film were colored in yellow and red and the

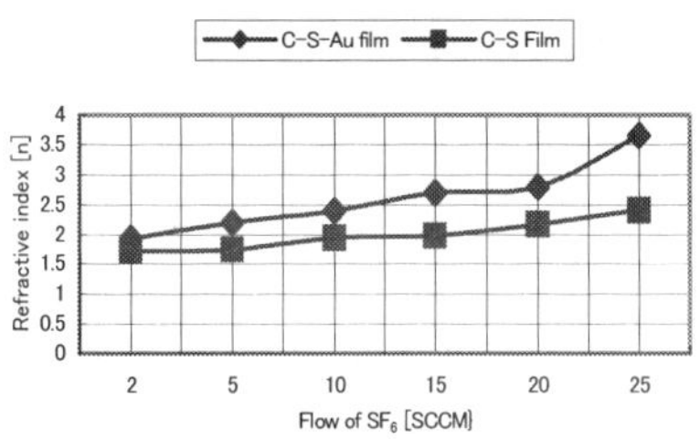

Fig.1 Refractive index of C-S & C-S-Au composite films (Ar. : 10 SCCM, CH_4: 10 SCCM, Pressure 0.1 Torr, Discharge duration 30 min).

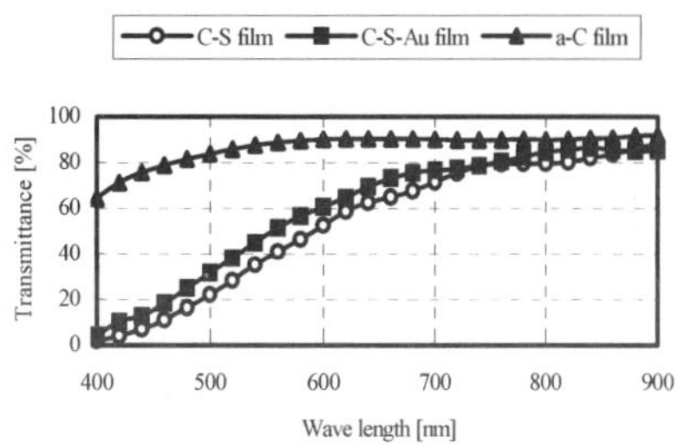

Fig.2 Optical transmittance of aC (600nm), C-S (1200) and C-S-Au (1100) film.

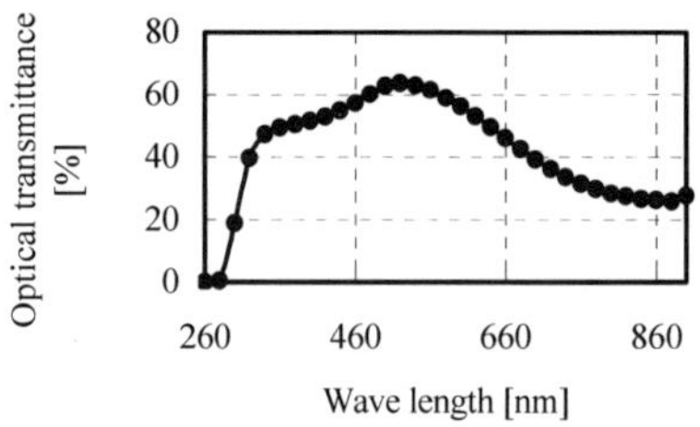

Fig.3 Optical Transmittance of Au film(Thickness 25 nm)

C-S-Au film showed absorption significantly at the shorter wavelength range than 400 nm. In Fig. 3, optical transmittance of gold film was measured. The Au film is transparent at the larger wavelength than 300 nm. Therefore the large absorption of C-S-Au film at the short wave length is referred to chemical bond between Au and S [11, 12]. To utilize the C-S-Au film for optical devises the deposition condition must be optimized to minimize the absorption.

Two-dimensional photonic crystal (2-D PC) was aimed to fabricate in this work as shown in Fig. 4, which will work at a visible light wavelength range of 500 nm. Where dot patterns have 500 nm pitch in a triangle structure. Therefore the gap distance between the dots is very small like as 200 nm so on.

Initially the etching rate of C-S-Au film by O_2 RIE plasma was measured as shown in Fig. 5. The etching was performed at 0.1 Torr, 10 SCCM of O_2 gas flow rate as a parameter of discharge power. The etching rate was very small. Therefore, to realize 2-D PC, it was concluded to adopt double layer resist system of evaporated aluminum (Al) and conventional resist. The resist must be high-resolution chemically amplified negative resist. The 2-D PC pattern of C-S-Au will be etched by O_2 RIE plasma through Al mask pattern formed by electron beam lithography. Where Al layer works as O_2 plasma etching stopper.

4.Conclusions

A new large refractive index material was fabricated by co-operation process of plasma CVD and sputtering with using CH_4 and SF_6 and Ar mixture gas and Au plate on the upper

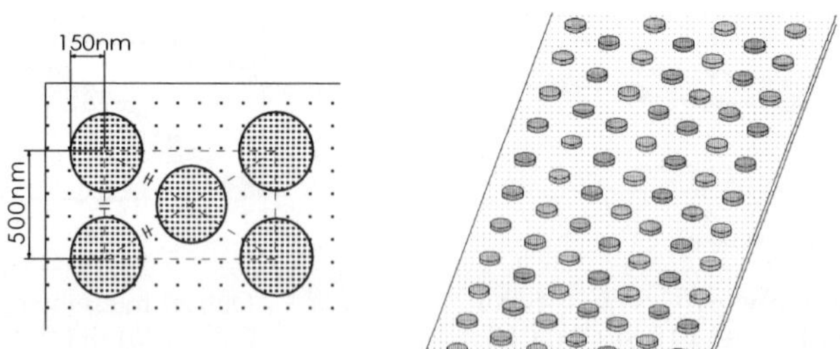

Fig.4 2-D PC pattern work at 500 nm.

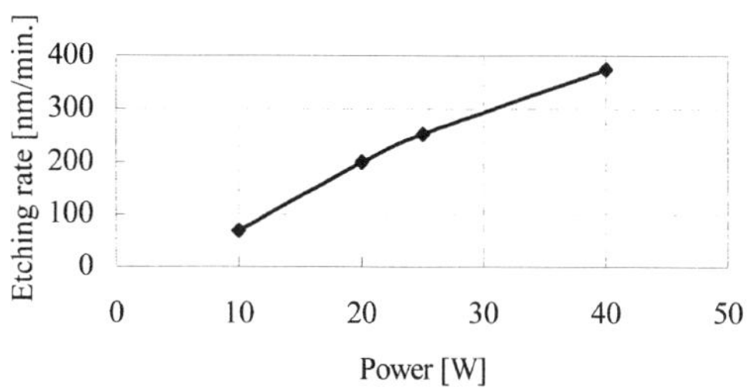

Fig.5 Etching rate as a function discharge power for C-S-Au film at 0.1 Torr and 10 SCCM of O_2.

electrode. C-S-Au composite film was formed and showed 3.7 as a refractive index. The value is suitable for the photonic crystal. The etching rate was measured to fabricate 2-D PC. Because of small etching rate, two layers resist systeBm of Al and resist were considered to be necessary to fabricate 2-D PC.

Acknowledgments

We would like to express sincere thanks to Prof. Hayakawa, director of Center of Co-operative Research in Advanced Science and Technology for the experiments.

References

[1] Hironori Sasaki, Kyoko Kotani, Hiroshi Wada, Takeshi Takamori and Takashi Ushikubo:*Oyo Buturi (in Japanese),* **69** (2000) 52.

[2] K. Wada and Lionel C. Kimerling:*Oyo Buturi (in Japanese),* **68** (1999) 1034.

[3] Eli Yablonovitch, *Phys. Rev. Lett.* **58**, (1987) 2059.

[4] J. D. Joannopoulos, R. D. Meade, and J. N. Winn, *Photonic Crystals,* Princeton University Press, (1995).

[5] S. Noda, K. Tomoda N. Yamamoto, and A. Chutinan, *Science*, **289** (2000) 604.

[6] M. Matsushita, Md. Abul Kashem and S. Morita, *IEICE Trans. on Electronics.,* **E83-C** (2000) 1134.

[7] M. Matsushita, Md. Abul Kashem and S. Morita, *Thin Solid Film,* **407** (2002) 50.

[8] Tsutomu Tukada, *Handbook on Plasma Material Science* Ed. By Kazuo Akashi, (1942) 492

[9] T. Inuzuka, *Diamond Thin Film (in Japanese)*, Kyoritsu Shuppan, Tokyo, Jan. 20 (1990).

[10] H. Biederman and L. Martinu, *Plasma Deposition, Treatment, and Etching of Polymers,* Ed. by R. d'Agostino, Academic Press, (1990) 269.

[11] K. Ishikawa, T. Isonaga, S. Wakita and Y. Suzuki, *Solid State Ionics,* **79** (1995) 60.

[12] S. Kera, H. Setoyama, K. Kimura, A. Iwasaki, K. K. Okudaira, Y. Harada and N. Ueno, *Surface Science,* **482-485** (2001) 1192.

Rapid Thermal Processing for Future Semiconductor Devices
H. Fukuda (Editor)

The LEVITOR 4000 system, Ultra-fast, Emissivity-independent, heating of substrates via heat conduction through thin gas layers

E.H.A. Granneman, V.I. Kuznetsov, A.B. Storm and H. Terhorst

ASM International nv, Jan van Eijcklaan 10, 3723 BC Bilthoven, The Netherlands
Phone:+31-30-2298411 Fax:+31-30-2298533 E-mail: Ernst.Granneman@asm.com

1.Introduction

During the last few years there has been an increasing interest in RTA systems that are not based on radiative heating [1-4, 6]. In these systems the wafer is placed in a resistively heated, isothermal, environment, where heating takes place through a combination of radiation and gas conduction. The driving forces for this trend are several: simple, thermocouple-based temperature control, no overshoot of wafer temperature, reduced dependence on wafer emissivity, low power consumption, etc. Further, as lamp-based systems have difficulties operating at temperatures < 400°C, specific furnace-based RTA systems are developed for low-temperature applications. In this paper the Levitor 4000 is discussed: an isothermal system, that can be used over the entire spectrum of RTA applications, i.e. from spike anneals (~1s, 1100°C) to low-temperature BEOL applications, i.e. annealing of ECD Cu and low-k dielectric films (>10s, < 400°C).

2.Basic Considerations

In isothermal systems, the wafer heat-up rate depends on the wall temperature, the distance between wafer and wall(s), and the gas ambient. In figure 1 a calculation is shown on the

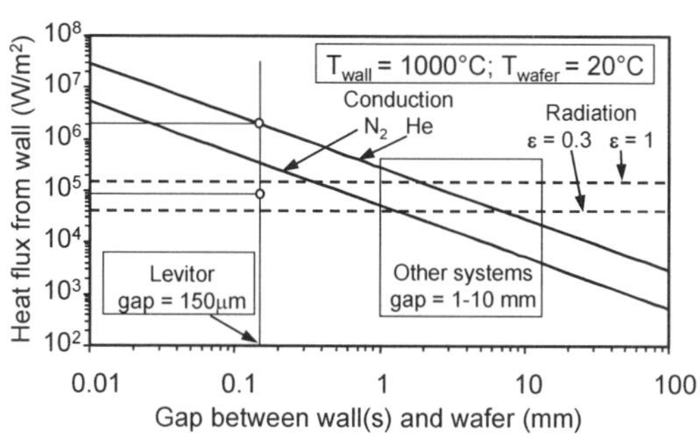

Fig.1 Comparison of radiative and conductive heating in isothermal systems. The assumption is that the wafer is placed in between two hot walls both kept at 1000°C.

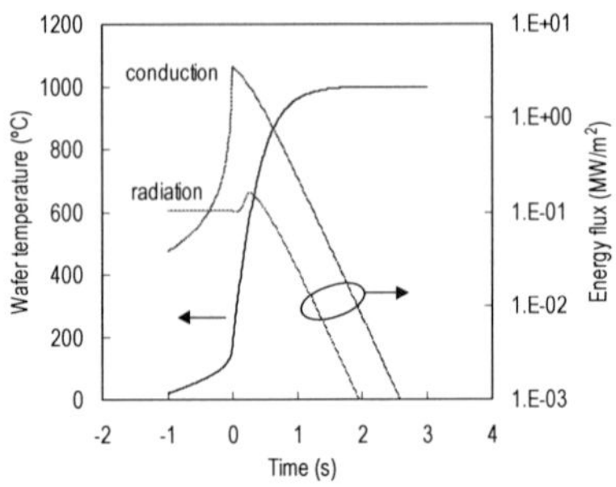

Fig.2 Wafer temperature and energy fluxes during wafer heat-up in helium to 1000°C. The gap between the wafer and both walls is 0.15 mm.

contributions of radiative and conductive heating, immediately after the wafer is placed in the system. As is clear from the figure, the gap between the walls and the wafer, and the gas ambient are critical in terms of the dominant heating mechanism. Systems with gaps >1mm [1-3] operate in a mixed radiation/conduction regime whereas the Levitor system [4,6], in which the gap between the wafer and the walls is 0.15 mm, operates in the conduction-dominated regime. In helium the fraction of energy transferred through conduction is ~ 95%.
In figure 2 the conductive and radiative contributions are shown all through heat-up of the wafer to the final (i.e. wall temperature). The heating turns out to be dominated by gas conduction throughout the entire heating process. This has important implications: heating through gas conduction, as takes place in systems with small gaps, is independent of wafer surface emissivity (device-type, patters, films, etc.). Further, very high heat-up rates are accomplished. In helium a maximum heat-up rate of ~ 900°C/s is realized; in nitrogen this is ~ 300°C/s

3.Levitor System

In the Levitor 4000 system, the wafer is inserted in between two blocks kept at constant temperature. Prior to loading the wafer, the blocks are at a typical distance of ~ 20 mm. Immediately after positioning the wafer in between the blocks, these are moved towards the wafer to a final and uniform wafer-block distance of 0.15 mm. During the entire wafer

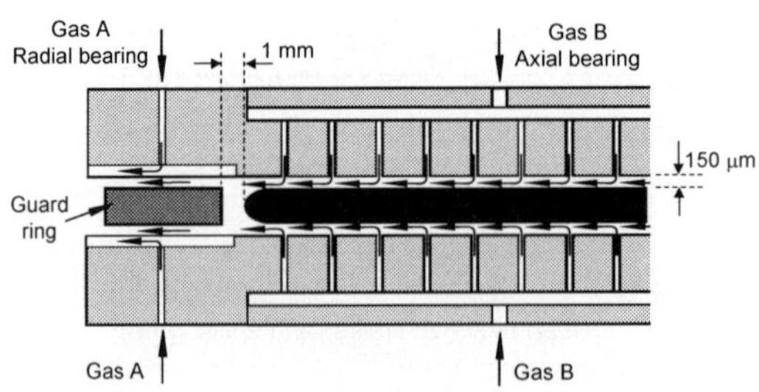

Fig.3 Basic lay-out Levitor 4000 system

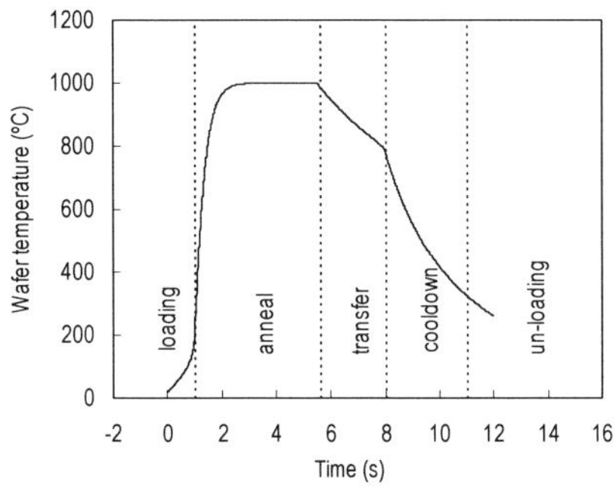

Fig.4 Typical time-temperature plot in the Levitor annealing system. The hot and cold blocks are closed during the 'anneal' and 'cool-down' times, respectively. In this example heat-up is done in helium, whereas cool-down takes place in nitrogen. During 'transfer', the wafer is transferred from the hot to the cold station. This takes typically 1.5-2.0 s.

loading sequence, gas is injected through narrow channels in both blocks, flowing radially outwards between wafer and blocks. The wafer floats stable on this gas bearing, see figure 3. Once the wafer is heated, the blocks are kept closed for as long as the recipe requires. When sufficient annealing has taken place, the blocks are opened and the wafer is taken out. To avoid the formation of slip, the so-called guard ring, see figure 3, is transported with the wafer. As this guard ring is heated to the same temperature as the wafer, it will shield the wafer edge from cooling down faster than the rest of the wafer during unloading. This eliminates the formation of crystal defects (slip). The wafer and the ring are transported to a cool-down station where the same principle is applied: wafer (and guard ring) are placed in between two water-cooled blocks that, upon loading of the wafer, are brought to a distance of 0.15 mm from the wafer surfaces. This results in a very efficient cool-down of the wafer, with rates of the same magnitude as are obtained during heat-up. Again, the dominant energy transfer mechanism is heat conduction through the thin gas layers. The overall process cycle is depicted schematically in figure 4.

Two variations of this set-up have been developed: in the first one, used for high-temperature anneals, both blocks are kept at the same temperature. As mentioned above, the wafer is heated rapidly to process temperature, with no overshoot. In the second version the blocks have different temperatures. The wafer temperature can be varied rapidly by either moving the wafer up or down in the thermal gradient, or by switching the type of gas above and below the wafer from a highly conductive type (He) to a less conductive type (N_2,), see figure 5.

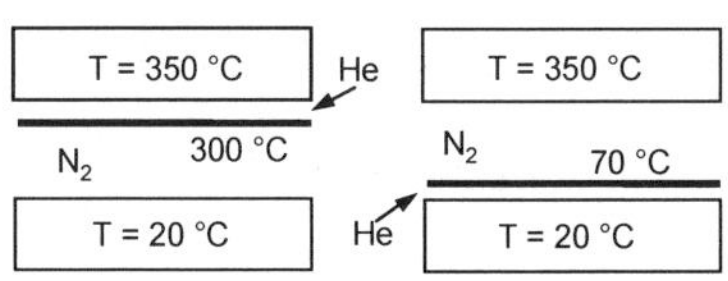

Fig.5 Levitor LT: two blocks with temperature gradient

Moving the wafer up and down in the thermal gradient is accomplished by varying the relative gas flows through the blocks on either side of the wafer. The advantage of this set-up is the relative simplicity, and the absence of a wafer transfer in between heating and cooling. The wafers are brought in cold, are heated up to the anneal temperature, and are cooled-down again without leaving the Levitor module.

4.Levitor 4000, Formation of Ultra-shallow junctions

To explore the performance of the system for the formation of Ultra-Shallow Junctions, various spike anneal processes were carried out on low-energy B implants. In one set of experiments the time the blocks are closed was kept constant at 1.3 s (= 'anneal' in figure 4) while the block temperature was varied. In the second set, the block temperature was kept constant at 1100 °C, and the time the blocks are closed was varied. In figure 6, the resulting junctions are shown. It is clear that the annealing process behaves as expected: the sheet resistance goes up at shorter block-closing times and at lower temperatures, while the junction depth is reduced. The combinations of sheet resistance and junction depth that are found in these experiments match the semi-theoretical limits for ultra-shallow junction formation as indicated by Huff et al [5].

It should be noted that for these experiments only 'conventional' Si substrates were used; no enhancement techniques such as the implantation of BF_2^+ ions, co-implantation of other ions and/or pre-amorphization are included. Such techniques result in even shallower junctions than the ones produced with the above-mentioned technique.

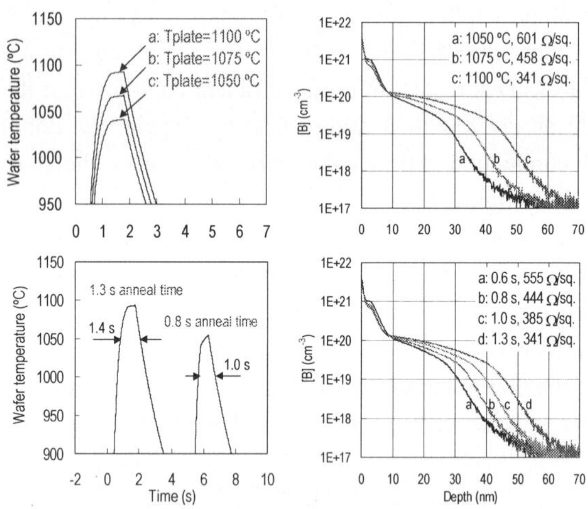

Fig.6 The formation of Ultra-shallow junctions with the Levitor 4000 system. In all cases 500 eV B^+ ions were implanted with a total dose of $1x10^{15}$ cm^{-2}. In the top two figures the block closing time ('anneal' time in figure 4) was kept constant at 1.3 s, while in the bottom two figures the block temperature was kept constant at 1100 °C.

Table 1 Backside-dependence of sheet resistance. 1 keV, 1.25E15 cm^{-2} B implants; anneal in He with blocks at 1100 °C, closed during 0.5 s.

Type Backside film	Rs (Ω/sq.)
160 nm Th.oxide, 90 nm Poly-Si	432
500 nm Th. oxide, 90 nm Poly-Si	437
300 nm Th. oxide, 130 nm Si3N4	441
420 nm PE-TEOS oxide	416

One of the advantages of conductive heating is the emissivity-independent heating of the wafers. To evaluate this, wafers with four different backside films were annealed in helium. The specific film combinations were chosen such that a reasonable variation in emissivity was obtained. The resulting sheet resistance values are summarized in table 1. It should be realized that the different wafers were processed in 'open loop' conditions; the temperature of the individual wafers was not measured, and no set point adaptations were applied to compensate for possible variations in emissivity. It can be concluded that the variation of sheet resistance for different backside films is a few percent at most. This indicates that the heating is indeed independent of emissivity for the films that were investigated.

5.Levitor-LT, Low-temperature anneals

In the low-temperature Levitor system [6], the wafer is inserted in between two blocks that are kept at different temperatures. Depending of the distance between the wafer and each of the blocks, and the gases present on each side, the wafer reaches a specific temperature. By varying the relative distances and gases, see figure 5, the wafer temperature can be made to vary between e.g. 60 and 300 °C, in either direction, in a matter of seconds, see figure 7. The (local) wafer temperature is measured in two ways: Firstly, thermocouples are pressed lightly against the top surface of the wafer. As these thermocouples enter the system through the hot block, the measured temperature will always be to high. At the same time the actual wafer temperature was determined by measuring the resistance of thin film temperature-dependent

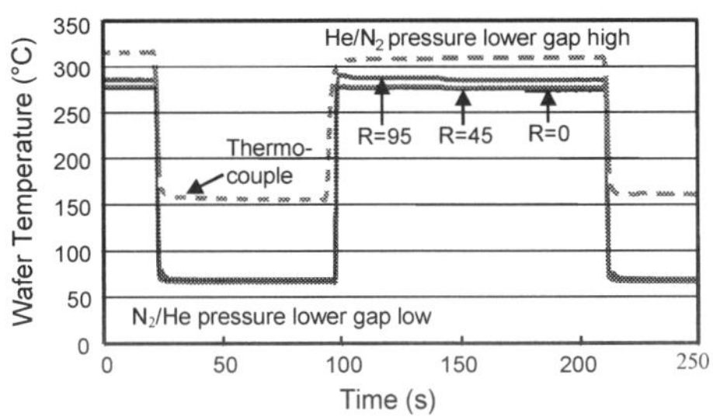

Fig.7 Wafer temperature (at three radii R) in system in which the position of the wafer in between two blocks is rapidly varied, as well as the gases on either side. The wafer reaches a high (low) temperature when the gas above (below) the wafer is helium, while nitrogen is present at the other side. Rapid temperature changes can are induced when the type of gas is switched, while at the same time the wafer position is shifted from close to the bottom block to close to the top block.

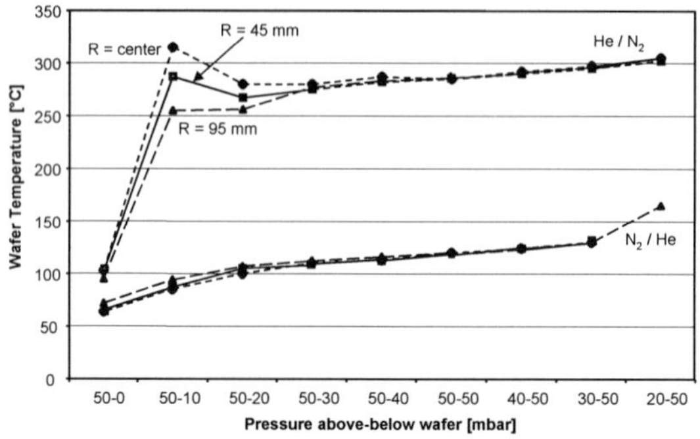

Fig.8 The temperature of the wafer as a function of the pressure above and below the wafer, and the type of gas present above and below the wafer. In principle, all points from the chart are accessible within seconds.

(Ti) resistors that are integrated on the wafer surface at radii 0, 45 and 95 mm. It is not completely clear whether the small temperature difference between r = 0 and 45 mm on the one hand and r = 95 mm on the other hand is caused by incorrect calibration of the Ti resistors, or by 'rest' bow being present in the wafer.

A particular benefit of this set up is that the gas that flows in the gas bearings above and below the wafer prevents diffusion of unwanted impurities from the outside inwards. Because of the relatively high flow, in combination with the narrow gap, the wafer is fully 'protected' while the blocks are closed (and the bearing gases are flowing). As in a typical process sequence, the wafer is loaded/unloaded on/from the cold plate, the blocks will always be closed when the wafer is hot, i.e. in the annealing phase. This is advantageous in e.g. Cu annealing: oxidation of the Cu film is completely avoided, despite the fact that the system is placed in air. Consequently, a gate valve separating the annealing module from the outside world is not required.

Experiments on Cu annealing, carried out with this module showed that ECD Cu films are fully annealed to bulk resistivity values in a wide regime of temperatures and annealing times (200-350°C, 15-60 s). More details on Cu annealing process results can be found in Granneman et al. [6].

Conclusions:

It has been demonstrated that wafers can be heated up and cooled down very rapidly by utilizing conduction heating (cooling), rather than radiation. When the wafer is placed in between two blocks kept at well-defined temperatures, at a distance of 0.15 mm from each of the blocks, energy transfer for the blocks to the wafer and vice versa through conduction is very fast. By floating the wafer in gas bearings the wafer does not touch the blocks, and is flattened by the forces exerted by the gas flows.

It has been demonstrated that this technology can be used for spike annealing processes, to produce advanced, ultra-shallow junctions. Further, the same principle can be applied towards low-temperature applications such as Cu annealing. In the latter case the two blocks are operated at different temperatures. By switching gases (He, N_2) above and below the wafer, and by varying the position of the wafer in between the two blocks, fast heat-up and cool-down is realized without removing the wafer from the module. It is to be expected that this technology will be applied in additional processes in the future.

References

[1] W.S. Yoo and A.S. Atanos, Proc. 7th Int. Conf. On Adv. Thermal Proc. of Semiconductors, Colorado Springs, Sept. 8-10, 1999, Edt. H. Kitayama, et al, page 259.

[2] Y. Senzaki, J. Sisson, R. Herring, J. Yoa, D.Teasdale and J. Ricketts, Int. Symp. On Rapid Thermal And other Short-time Proc. Technol., ECS 2001 Proc. Abstr.563

[3] W.S. Yoo, T. Yamazaki, T. Uchino, Sol. St. Technol. Vol.6, 2001, page 152.

[4] V.I. Kuznetsov, A.B. Storm, G.J. Snijders, C. de Ridder, T.A.M. Ruijl, J.C.G. v.d. Sanden and E.H.A. Granneman. Int. Symp. On Rapid Thermal and other Short-time Proc. Technol. ECS Proc. 2000-9, page 401.

[5] H.R. Huff, G.A. Brown and L.A. Larson, Int. Symp. On Rapid Thermal And other Short-time Proc. Technol., ECS 2001 Proc. , PV 2001-2, ULSI Process Integration, eds. C. Claes, F. Gonzalez, J. Murota and K. Saraswat, page 223

[6] E.H.A. Granneman, V.I. Kuznetsov, A.B. Storm, J.B. Post, T. Ritzdorf, R.A. Weaver, R. Baskaran. Int. Symp. on Rapid Thermal And other Short-time Proc. Technol., ECS 2001 Abstr.559.

Rapid Thermal Processing for Future Semiconductor Devices
H. Fukuda (Editor)

Steady and Transient Gas Flow Simulation of SiGe Vertical Reactor

Akihiro Miyauchi[1] , Hirohisa Yamazaki[2], Yasuhiro Inokuchi[2] and Yasuo Kunii [2]

[1] Hitachi Research Laboratory, Hitachi Ltd., 7-1-1 Ohmika, Hitachi, Ibaraki, 319-1292 Japan
Phone:+81-294-52-5111 Fax:+81-294-52-7622 E-mail: amiyauch@hrl.hitachi.co.jp
[2] Toyama Works, Hitachi Kokusai Electric Inc., 2-1 Yasuuchi, Yatsuo, nei-gun, Toyama, 939-2393 Japan
Phone:+81-76-455-9111 Fax:+81-76-455-9165 E-mail: kunii.yasuo@h-kokusai.com

1.Introduction

SiGe is an attractive material for advanced high-speed devices. SiGe has been applied for a base region of hetero-junction bipolar transistors (HBTs) [1]. SiGe is also useful for CMOS devices, such as SiGe gate structures, elevated source/drain structures and strained Si channel devices. SiGe films have been formed by single wafer equipment because the single wafer equipments have achieved precise process control.

A batch system is able to treat a large number of wafers at a time. However, gas flow states are complicated and it has been unknown that precise and abrupt profiles could be achieved. We analyzed the steady and transient gas flows in the batch reactor and found that the process conditions for sub-nanometer scale process control was available by the vertical batch reactor.

2.Simulation method

A three-dimensional (3D) computer aided design (CAD) data was made based on the actual batch reactor as shown in Fig. 1. The conservative equations of mass, momentum and energy were calculated in each cell of the 3D CAD. The number of cells was 270,000. The temperatures, pressures and gas velocities were obtained in each cell. Binary diffusion coefficients between source gases and hydrogen carrier gas were calculated using eq. 1 and 2.

$$D^0_{i',j'} = 4.3E-3\frac{273^{3/2}\left(1/M_{i'} + 1/M_{j'}\right)^{1/2}}{P\left(V_{i'}^{1/3} + V_{j'}^{1/3}\right)^2} \tag{1}$$

$$D_{i',j'} = D^0_{i',j'}\left(\frac{T}{T_0}\right)^m\left(\frac{P_0}{P}\right) \tag{2}$$

where M*i* is mass weight, V*i* is mole volume, T is temperature, P is pressure and m is 1.833.

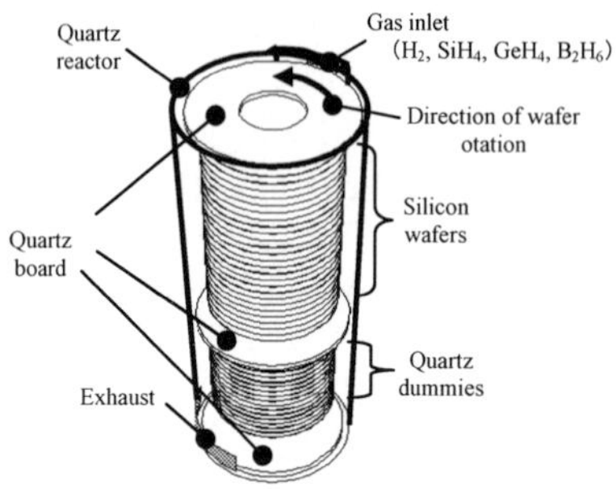

Fig.1 Schematic view of the simulated reactor.

Table I Simulated process conditions

Parameters	
Mass fraction	$H_2:SiH_4:GeH_4:B_2H_6$ $=0.77:0.2:0.02:0.01$
Quartz wall temp	500°C
Wafer rotation	1 rpm
Reactor pressure	300 Pa -1 atm

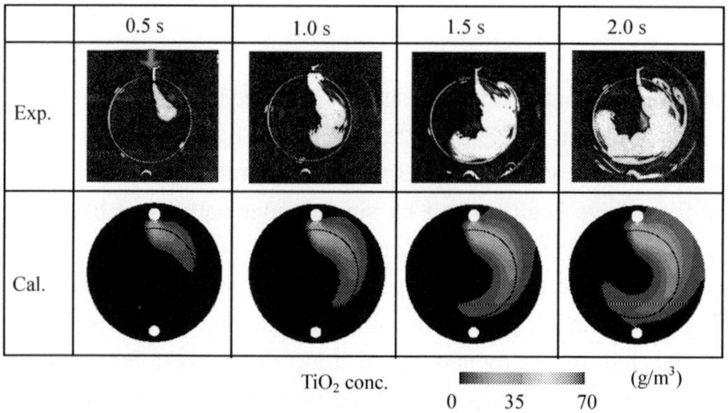

Fig.2 Measured and calculated gas flow states after TiO_2 injection.

3.Results

Gas has viscosity so that wafer rotation affects on the gas flow states. We investigated the method of gas flow simulation comparing the visualized gas flow data. The viscosity and the binary diffusion coefficients were calculated exactly by a kinetic theory using Lennard-Jones potential. The fine mesh structure was applied in the boundary layer.

Fig. 2 shows the comparison of the experimental and calculated results. The experimental results were obtained by injection of the TiO_2 and video recording. The transient behavior of TiO_2 smokes was well analyzed by using the fine mesh structure and viscosity calculations. This model was expanded to the vertical reactor to predict the transient behavior of B_2H_6 gas.

Fig. 3 shows steady gas flow states as a function of pressures in the SiGe reactor. The gas streams were straight from inlet to exhaust at the pressures below 12 kPa. The streamlines were disturbed by the wafer rotation above 50 kPa.

Fig. 4 shows the steady gas flow states between the wafers. There were rotational motions at 12 kPa but one directional flows were formed at 0.3 kPa.

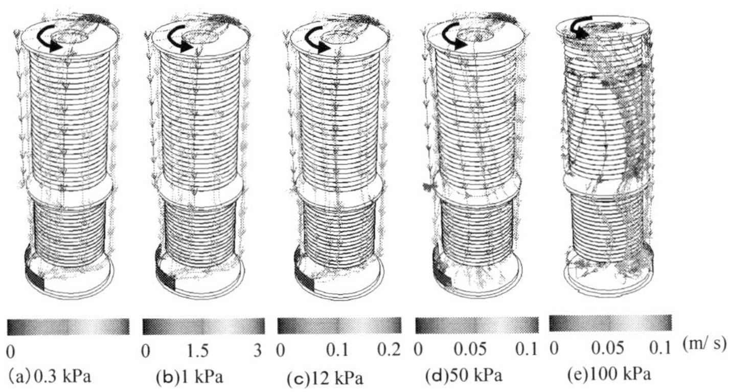

Fig.3 Simulated steady state gas flows in the reactor as a function of pressures.

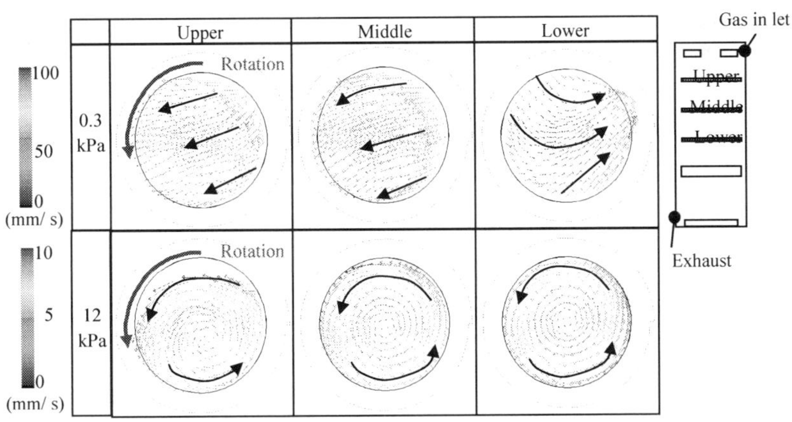

Fig.4 Steady state gas flows between the rotating wafers as a function of wafer locations and pressures.

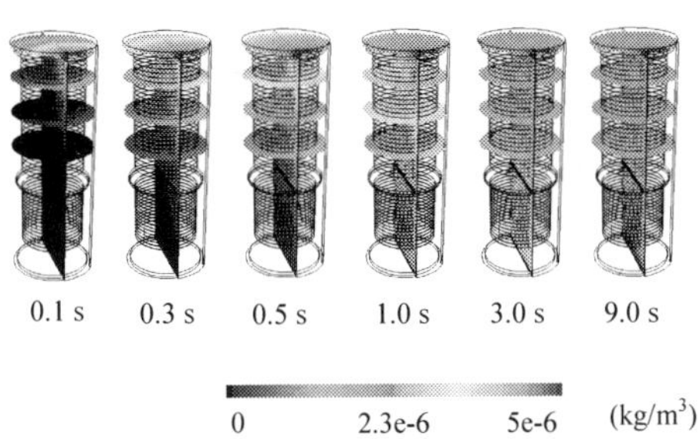

Fig.5 Transient behavior of B_2H_6 distribution after introducing the B_2H_6 gas.

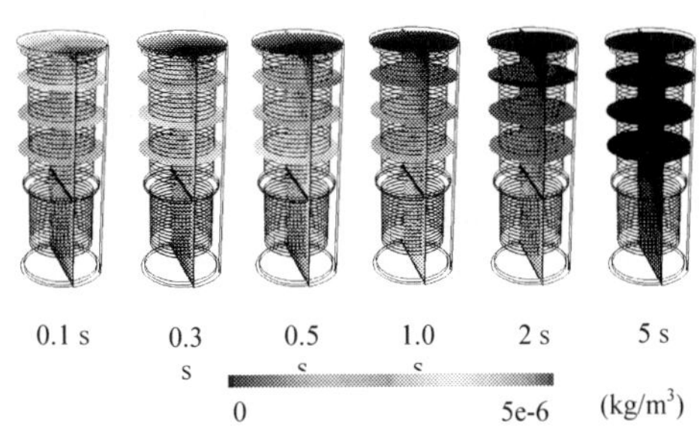

Fig.6 Transient behavior of B_2H_6 distribution after stopping the B_2H_6 gas.

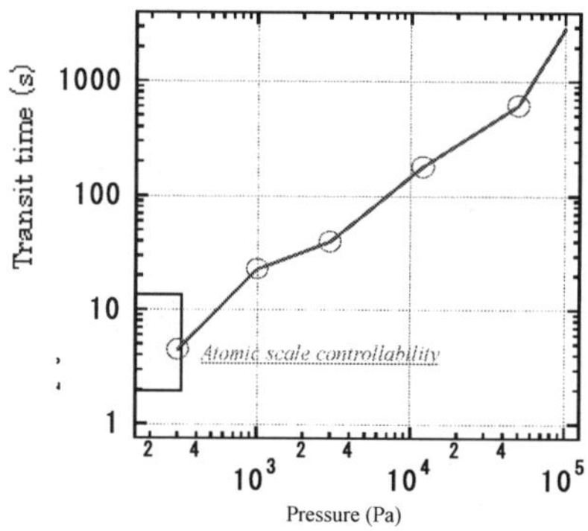

Fig.7 Transient behavior of normalized B_2H_6 concentrations as a function of pressures.

Fig. 5 shows the results of transient behavior of B_2H_6 gas after introducing the B_2H_6 gas. B_2H_6 gas was spread from the top of the reactor, and went down to the bottom of the reactor. B_2H_6 gas was filled uniformly after 9 seconds.

Fig. 6 shows the transient behavior of B_2H_6 gas after stopping the B_2H_6 supply. The B_2H_6 gas was purged out from the top of the reactor, and evacuated at the outlet that was located in the bottom of the reactor. B_2H_6 gas was purged after 5 seconds.

Fig. 7 shows the transient time to fill and purge-out the B_2H_6 gas as a function of the pressures. Assuming the condition of sub-nanometer thickness control of SiGe base formation, B_2H_6 gas concentration must be controlled in less than 5 seconds. This rapid process was achieved by reducing the pressures to below 320 Pa.

4.Conclusions

The steady and transient gas flow states in the SiGe vertical reactor were analyzed based on the visualized gas flow data. The precise control of HBT base process was achieved by reducing the pressures to below 320 Pa.

Acknowledgments

We would like to thank to Drs. Y. Inoue and Y. Kobayashi for their advice and support.

References

[1] K. Washio, et al., IEDM Tech. Dig., 557 (1999).

Rapid Thermal Processing for Future Semiconductor Devices
H. Fukuda (Editor)

The Short-period $(Si_{14}/Ge_1)_{20}$ and $(Si_{28}/Ge_2)_{10}$ superlattices as Buffer Layers for the Growth of $Si_{0.75}Ge_{0.25}$ Alloy Layers

M.M. Rahman, K. Kurumatani, T. Tambo, and C. Tatsuyama

Department of Electrical and Electronic Engineering, Faculty of Engineering, Toyama University, 3190-Gofuku, Toyama 930-8555, Japan
Phone and Fax: +81-76-445-6728, E mail: mmizanurr@hotmail.com

1.Introduction

Si and Ge are very prospective and widely used semiconductors in electronic and optoelectronic device fabrications. 4.2% lattice mismatch between Si and Ge limits its epitaxial growth. However, this lattice mismatch can be accommodated by a finite degree of lattice distortion, which is called a pseudomorphic strained layer. This pesudomorphicity maintains up to a thickness called critical thickness h_c. Several models have been reported on the critical thickness [1-4]. Above the critical thickness strain energy becomes too high to maintain more layers in stress. As a result strained layers goes to relaxation introducing high density of misfit dislocations. Extended part of misfit dislocations can come up to the upper surface, which is called threading dislocations. Threading dislocations in the active device act as leakage current channels or scattering centers for carriers. Due to these stray affects, the devices with defects lost its reliability and perfections, and drastically reduce the performance. For this reason, research on $Si_{1-x}Ge_x$ has got tremendous attraction. To achieve dislocation free $Si_{1-x}Ge_x$ layers, various types of sacrificial buffer layers such as thick $Si_{1-y}Ge_y$ $(x \geq y)$[5], graded SiGe, step graded $Si_{1-x}Ge_x$ buffers [6-10] and $Si_{1-x}Ge_x/Si$ superlattices [11-13] have been used. In our previous reports [14-16], we have used short-period superlattices with various steps for the growth of $Si_{0.75}Ge_{0.25}$ alloy layers and observed dramatic reduction in surface roughness.

For device fabrication, the thinner buffer layer with lower thermal budget is essential. In this paper we report on the effect of growth temperature of $(Si_{14}/Ge_1)_{20}$ and $(Si_{28}/Ge_2)_{10}$ short-period superlattice (SSL) buffers on the growth of 2000-Å -$Si_{0.75}Ge_{0.25}$ alloy layers.

2.Experimental details

In MBE system (ANELVA-620s), 2000-Å-$Si_{0.75}Ge_{0.25}$ alloy layers were grown using $(Si_{14}/Ge_1)_{20}$ and $(Si_{28}/Ge_2)_{10}$ SSL buffer layers on Si(001) substrates and we named these two series of samples as type-A and type-B samples, respectively. Details of sample preparation and MBE system are discussed in elsewhere [15]. Here, subscripts in the SSLs (with Si 14

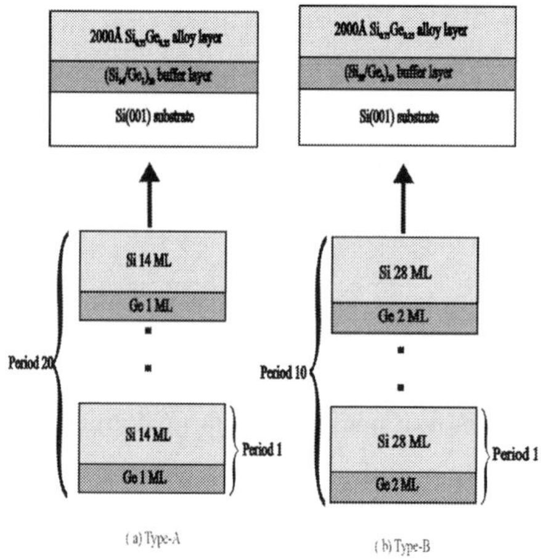

Fig.1 Schematic diagram of the two types of samples and their corresponding buffer layers.

and 28, with Ge 1 and 2) inside the first bracket denote the monolayers of Si and Ge, respectively, and the subscripts (20,10) outside the bracket denote the period of (Si/Ge) bilayers. Buffer layers of the two types of samples were grown at different temperature from 300 to 450°C i.e. within the low-temperature MBE limit. On this buffer compositionally uniform 2000-Å -$Si_{0.75}Ge_{0.25}$ alloy layers were grown at 500 °C. Schematic diagram of two types of samples are shown in Fig.1.

Residual strain in the alloy layers was estimated from the XRD data using Cu Kα x-rays. The surface morphology was observed by atomic force microscope (AFM) and the surface roughness was characterized by root-mean-squared (rms) surface roughness of the AFM images. Orientation of dislocations are investigated by cross-sectional transmission electron microscopy (XTEM).

3.Result and discussion

X-ray diffraction (XRD) observation

X-ray diffraction data of all the samples showed very clear Si(004) and $Si_{0.75}Ge_{0.25}$(004) alloy peaks (data is not shown). We have chosen the alloy layer thickness above the critical thickness according to some well-known reports [1,2]. Residual strain $\varepsilon_{\perp}$, perpendicular to the growth plane in the alloy layers was estimated from XRD data using procedure discussed in [14]. Estimated residual strain is plotted in Fig. 2 as a function of the growth temperature of the buffer layers. In A-type samples with Ge-1-monolayer mode SSL, it is seen that the residual strain decreases monotonically with decreasing of the growth temperature of the buffer layers and reached a value of about –0.08% at 300°C. Lower growth temperature of buffer layers may have better layered structures, because Ge segregation is lower for lower growth temperature. In our previous study, we have observed that substantial intermixing

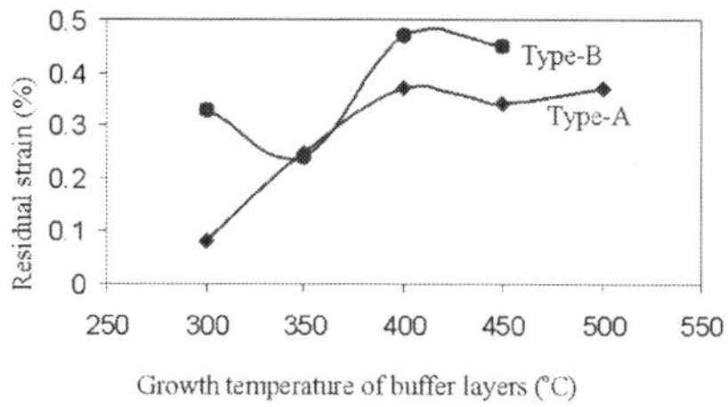

Fig.2 Residual strain of type-A and type-B samples as a function of the growth temperature of the buffer layers.

occurs at the interface of the Si-Ge SSL prepared at 500°C[17]. But Ge surface segregation at 500°C has been found to be suppressed with an increase of the Si deposition rate [18,19].

It is known there are upper limits on the number of strained-layer regions, which can be accommodated elastically on a given substrate [20]. For Ge grown epitaxially on Si, the maximum number of Ge MLs can be deposited is six [21]. But in the present experiment we have used only one ML and two MLs mode deposition. Another point is worth noticeable; when Si is deposited at lower temperature it introduces a large number of point defects. The strain in the SSL layers can deflect the dislocations and point defects can capture threading dislocations. As a result the upper layers not only become relaxed but also become smooth. The residual strain in B-type samples with Ge-2-monolayer mode in SSL layers seems to show decreasing trend, but it is higher than that of A-type samples. The reason of higher residual strain of this type of samples is not clear at this time.

Atomic Force Microscopy (AFM) Observation

AFM images of some samples of both types of samples are shown in Fig. 3, where 2(a-b) are samples (A-type) with SSL buffers grown at 450 and 300 °C, respectively, 2(c-d) are the samples (B-type) with buffers grown at 450 and 300°C, respectively. All the images show the cross hatch patterns, which is related to the underlying grid of misfit dislocations. From the figure, it is seen that the height of troughs and crests of cross hatches is low.

Root-mean-squared (rms) surface roughness of all the samples is shown in Fig. 4 as a function of growth temperature of the buffer layers. Without SSL buffer layer, the rms roughness of 2000-Å-$Si_{0.75}Ge_{0.25}$ alloy is high (about 40 Å, data is not shown here), whereas samples with SSL buffer layer show dramatic reduction in roughness. From Fig. 4, it is seen that the roughness of type-A samples is independent of the growth temperature of SSL buffer layers and gives about a constant value (~12 Å). As the strain energy increases with decreasing of the growth temperature, which can effectively deflect the threading dislocation, may be the cause of the constant rms roughness and smooth surface of the alloy layers. In the case of type-B samples rms roughness pattern is as like type-A samples but its roughness value is higher that that of type-A samples. Two-monolayer mode Ge deposition may cause island like pattern in the SSL structure. As a result, it may go to partial relaxation and ultimately can not deflect threading dislocations. So roughness increases. Some spots on the type –B samples is the evidence of the unfiltered threading dislocations.

Cross-sectional Transmission Electron Microscopy (XTEM) Observation

Some cross-sectional transmission electron microscopy (XTEM) images of type-A samples are shown in Fig. 5, where 2000 Å $Si_{0.75}Ge_{0.25}$ alloy layer with (a) 500°C-grown, (b) 450°C-grown and (c) 400°C-grown SSL buffer layers. From the figure, it is seen that the upper alloy layers of all the samples are clear and free from dislocations. A clear upper alloy is seen here, which is consistent with the lower roughness of AFM images. Different types of dislocations are seen in the images with different temperature growth buffers. In Fig. 5(a), dislocation half loops are seen in the substrate region. Dislocations are deflected from the strained layer superlattices. Pinning of dislocations in Si substrate has been explained by LaGoues et al.[22]. V-shaped dislocations (Fig. 5c) also shows the filtering of dislocation.

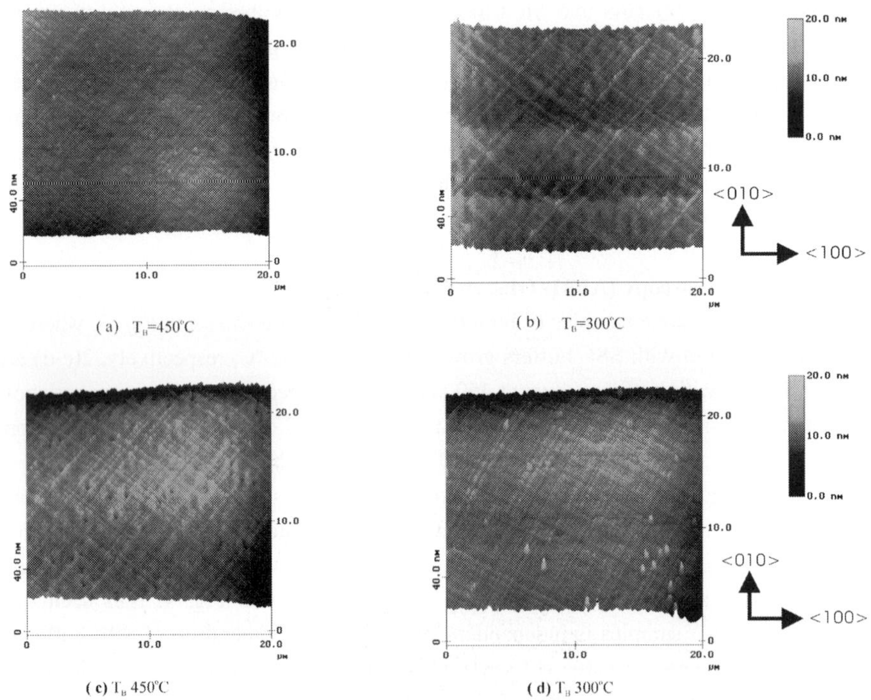

Fig.3 AFM image (20 μm x 20 μm) of the $Si_{0.75}Ge_{0.25}$ alloy layers with (a) 450°C-grown $(Si_{14}/Ge_{1})_{20}$ SSL, (b) 300°C-grown $(Si_{14}/Ge_{1})_{20}$ SSL, (c) 450°Cgrown $(Si_{28}/Ge_{2})_{10}$ and (d) 300°C-grown $(Si_{28}/Ge_{2})_{10}$ SSL buffer layers. All the alloy layers were grown at 500°C.

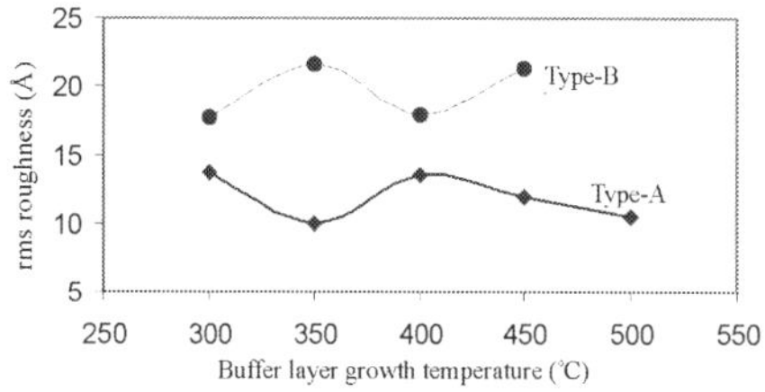

Fig.4 Root-mean-squared (rms) surface roughness of type-A and type-B samples as a function of the growth temperature of their respective buffer layers. In all the samples the alloy layers were grown at 500 °C

4.Conclusion

2000 Å $Si_{0.75}Ge_{0.25}$ alloy layers were grown at 500 °C using $(Si_{14}/Ge_{1})_{20}$ and $(Si_{28}/Ge_{2})_{10}$ SSL as buffer layers, where buffer layers were grown at different temperatures from 300-450 °C. In the first buffers, Ge was deposited as Ge-1-monolayer mode and in the second buffers, it was Ge-2-monolayer mode. Effects of growth temperature of these two buffer layers on the top alloy layers were characterized. From XRD data it was seen that the residual strain decreases with decrease of growth temperature of Ge-1-monolayer mode deposited SSL layers and reached a lower value of about –0.08%. The samples with these buffers also

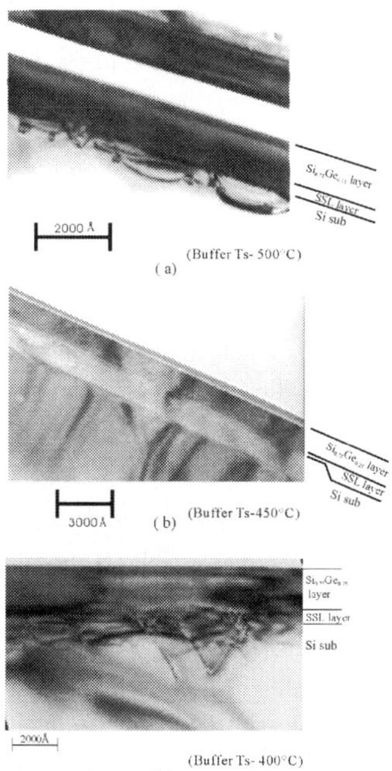

Fig.5 XTEM micrographs of samples 2000Å -thick $Si_{0.75}Ge_{0.25}$ alloy layers with (a) 500°C grown SSL, (2) 450°C grown SSL and (c) 400°C grown SSL buffer layers.

showed a smooth surface with about a constant rms roughness (~12 Å) irrespective of growth temperature of the SSL buffer layers. The samples with Ge-2-monolayer mode SSL buffers showed higher residual strain and higher rms roughness than those of the Ge-1-monolayer mode samples. XTEM image of the sample with Ge-1-monolayer mode SSL buffer showed that the top alloy layer is free from dislocations and all the dislocations are in both SSL buffer and substrate. So, we can say that the short-period $(Si_{15}/Ge_{1})_{20}$ SL buffer grown at 300°C is better for the growth of relaxed and smooth $Si_{0.75}Ge_{0.25}$ alloy layers.

Acknowledgement:

Authors would like to express sincere thanks to Dr. Matsuda of Material Science department, Toyama University for his kind help to take XTEM micrographs.

References

[1] J.W. Mathews and E.A. Blakeslee, J. Cryst. Growth. 27(1974)118.

[2] R. People and J.C. Bean, Appl. Phys. Lett. 47(1985)322.

[3] R. People and J.C. Bean, Appl. Phys. Lett. 49(1986)229.

[4] K. Shintani, and K. Fujita, J. Appl. Phys. 75(1994)7842.

[5] E. Kasper, and F. Schaffler, In *Strained Layer Superlattices: Material Science and technology, Semiconductor and Seimetals,* 33, T.P. Pearsall, ed., San Diego, CA: Academic Press, pp.223-311.

[6] E.A. Fitzgerald, Y.H. Xie, M.L. Green, D. Bransen, A.R. Kortan, J. Michel, Y.J. Mill, and B.E. Weier, Appl. Phys. Lett. 59, 811(1991).

[7] C.G. Tuppen, C.G. Gibbings, and M. Hockly, Proc. Mat. Res. Soc. 220(1991)187.

[8] Z. Tongyi, Phys. Stat. Sol. (a) 148(1995)175.

[9] P.M. Mooney, J.L. Jordan-Sweet, J.O. Chu, F.K. LeGoues, Appl. Phys. Lett. 66(1995)3642.

[10] J.H. Li, Koppensteiner, E. Bauer, M. Honisch, H.J. Herjog, and F. Schaffler, Appl. Phys. Lett. 67(1995)223.

[11] Z. Liliental-Weber, E.R. Washburn, T.Y. Liu, and H. Kroemer, Proc. Mat. Res. Soc. 91(1987)91.

[12] G.H. Oleson, M.S. Abrahams, G.J. Buicchi, and T.J. Zamerowski, J. Appl. Phys. 46(1975)1643.

[13] J.M. Hartmann, B. Gallas, J. Zhang, J.J. Harris, and B.A. Joyce, J. Appl. Phys. 86(1999)845.

[14] T. Obata, K. Kameda, H. Ueba, and C. Tatsuyama, J. Appl. Phys. 81, 199(1997).

[15] M.M. Rahman, H. Matada, T. Tambo, and C. Tatsuyama, J. Appl. Phys. 90, 202(2001).

[16] M.M. Rahman, H. Matada, T. Tambo, and C. Tatsuyama, Appl. Surf. Sci. 175, 6(2001).

[17] T. Tamagawa, T. Shintani, H. Ueba, K. Nakagawa, M. Miyao, and C. Tatsuyama, Thin Silid Films, 237(1994)282.

[18] K. Nakagawa, and M. Miyao, J. Appl. Phys. 69(1991)3058.

[19] K. Nakagawa, M. Miyao, and Y. Shiraki, Thin Solid Films, 183(1989)315.

[20] R. Hull, J.C. Bean, F. Cardeira, A.T. Fiory, and J.M. Gibson, Appl. Phys. Lett. 49(1986)1287.

[21] J. Bevk, J.P. Mannaerts, L.C. Feldman, B.A. Davidson, and A. Ourmazd, Appl. Phys. Lett. 49(1986)286.

[22] F.K. Legoues, B.S. Meyerson and J.F. Morar, Phys. Rev. Lett. 66, 2903(1991).

Rapid Thermal Processing for Future Semiconductor Devices
H. Fukuda (Editor)

Si Epitaxial Growth on the Atomic-Order Nitrided Si(100) Surface in SiH_4 Reaction

Youngcheon Jeong, Masao Sakuraba, Takashi Matsuura and Junichi Murota*
Laboratory for Electronic Intelligent Systems, Research Institute of Electrical Communication, Tohoku University
2-1-1 Katahira, Aoba-ku, Sendai 980-8577, Japan
Phone&Fax: +81-22-217-5548 E-mail: murota@riec.tohoku.ac.jp

1.Introduction

Atomically controlled processing such as layer-by-layer growth is attractive for creating novel nano-structures and superlattice structures in Si technology. Silicon nitride films under atomic-order thickness control can be applied to tunneling barrier and very thin gate insulator. In our previous work [1,2], atomic-layer nitridation of Si(100) and possibility of subsequent Si epitaxy at lower temperatures were demonstrated using an ultraclean low-pressure chemical vapor deposition (CVD) system. In the present work, Si epitaxial growth conditions on the atomic-order nitrided Si(100) surface has been clarified using the low-pressure CVD system. Furthermore, the thermal stability of the N atoms in the N delta-doped Si films has been investigated.

2.Experiment

The ultraclean hot-wall low-pressure CVD system [3] used for Si epitaxial growth on the nitrided Si surface formed by NH_3 is schematically shown in Fig. 1. The system was made ultrahigh vacuum compatible with gate valves and a turbo molecular pump. To minimize air contamination into the reactor during wafer loading and unloading, an N_2 purged transfer chamber was combined at the reactor inlet. The substrates used were the p-type Si(100) wafers of 2-20 Ω cm with a mirror-polished surface. They were cleaned in several cycles in a 4 : 1 mixed solution of 98%-H_2SO_4 and 30%-H_2O_2, and rinsed with deionized water. Then,

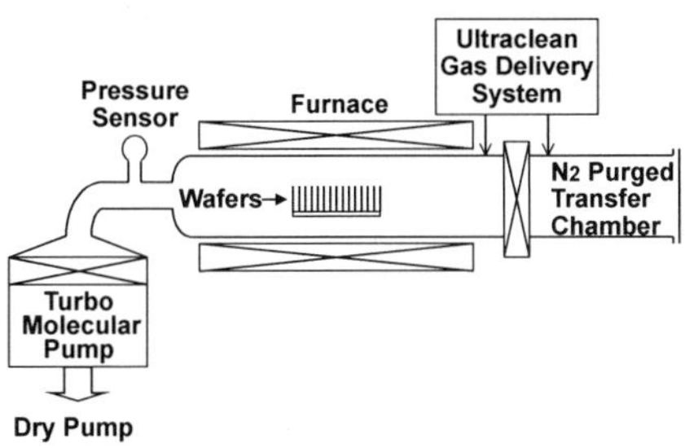

Fig.1 Schematic diagram of the ultraclean hot-wall low-pressure CVD system.

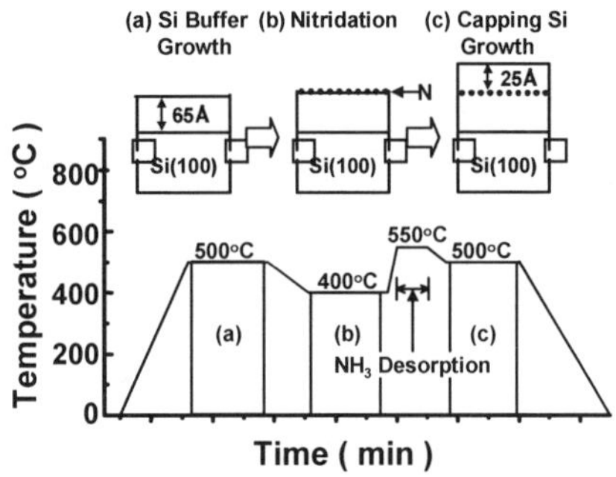

Fig.2 Typical process sequence for deposition of the N delta-doped Si films with the ultraclean hot-wall low-pressure CVD system.

just before wafer loading, they were dipped into a 2%-HF solution followed by the deionized water rinse for removing native oxide. The surface structure and the crystallinity of the deposited film were evaluated by a reflection high-energy electron diffraction (RHEED).

The typical process sequence for the deposition of the N delta-doped Si film was shown in Fig. 2. At first, a 65Å-thick Si buffer layer was grown on the Si(100) substrates at 500°C with a SiH_4 pressure of 25 Pa for 20 min in order to prepare an atomically flat and contamination-free Si surface (Fig. 2(a)). Atomic-order nitridation of the Si surface was performed by NH_3 exposure at the pressure of 124 Pa at 400°C [4] (Fig. 2(b)). Thereafter, in order to reduce residual NH_3, H_2 was purged for 30min at 550 °C. Then, SiH_4 exposure was carried out for 20-60 min at the pressure 25 Pa at 500°C for capping Si growth (Fig. 2(c)).

In order to obtain the depth profile of the N concentration for the N delta-doped Si film, XPS measurement and chemical etching for the N delta-doped Si film were repeated. Here, the chemical etching was done by dipping the sample in a 4:1 mixed solution of 98%-H_2SO_4 and 30%-H_2O_2 for 15min and subsequently in a 2%-HF solution followed by the deionized water rinse shown in Fig. 3. The etching rate of Si(100) was 4.2Å/cycle in average, which was measured by atomic-force microscope (AFM) using a partially SiO_2-covered Si(100) substrate. Substituting N 1s intensities measured by XPS at each etching cycle to equation (1), we can determine the average N concentration in the etched region with taking the photoelectron escape depth (λ) in Si into consideration and the overall conversion factor (β) which was calibrated by standard atomic-layer order nitrided sample.

$$N_i = \frac{I_i - I_{i+1}\exp(-\frac{x_{i+1}-x_i}{\lambda})}{\beta\lambda\left\{1-\exp(-\frac{x_{i+1}-x_i}{\lambda})\right\}} \qquad (1)$$

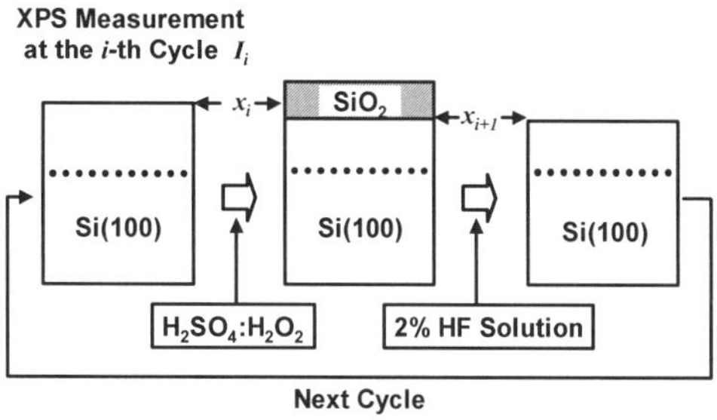

Fig.3 Typical measurement method for depth profile of the N concentration in the atomic-layer order N delta-doped Si films.

3.Results and Discussion

The Si Deposition on the Atomic-Order Nitrided Si(100) Surface

RHEED patterns taken from the N delta-doped Si film are shown in Fig. 4. Clear streaks and Kikuchi lines originated from a flat epitaxial Si(100) surface are observed even in the case at the initial surface N amount as high as $3.0\times10^{14}cm^{-2}$ (Fig 4(a)), although it is changed into the halo pattern originated from amorphous in the case at the initial surface N amount of $6.0\times10^{14}cm^{-2}$ (Fig 4(b)). Furthermore, clear steaks are also observed for Si films with four N delta-doped regions (Fig. 4 (c)). This result indicates the possibility of epitaxial superlattice formation with Si nitride and Si.

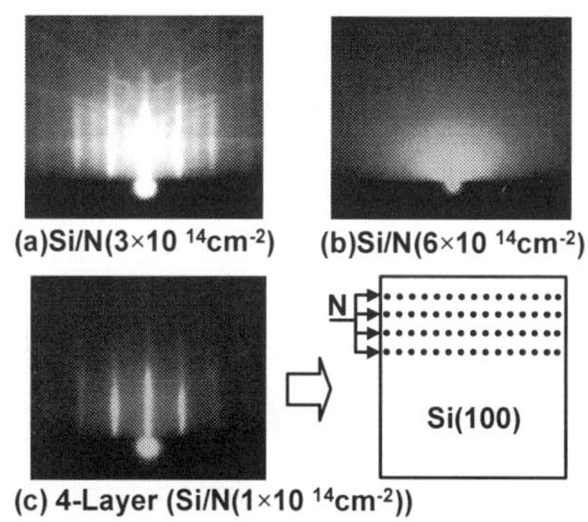

Fig.4 RHEED patterns taken from the [011] azimuth for the N delta-doped Si film, in which the capping Si thickness is about 25Å. The initial surface N amount on the buffer Si surface is (a) $3.0\times10^{14}cm^{-2}$, and (b) $6.0\times10^{14}cm^{-2}$. (c) the Si films with four N delta-doped regions separated by 25Å-thick Si spacer layer and the initial surface N amount is $1.0\times10^{14}cm^{-2}$ for each region.

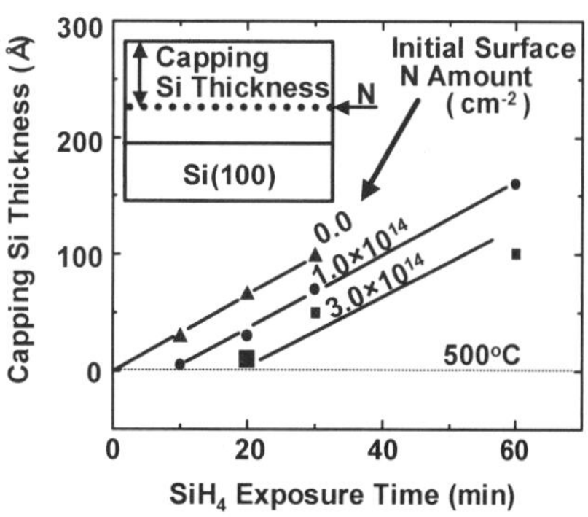

Fig.5 SiH_4 exposure time dependence of the capping Si thickness on the atomic-order nitrided Si(100) surface at 500°C. The SiH_4 pressure is 25 Pa.

The SiH_4 exposure time dependence of the deposited capping Si thickness on the atomic-order nitrided Si(100) surface is shown in Fig. 5. An incubation period of Si deposition on the atomic-order nitrided surface is found, while the deposition rate after the incubation period is almost the same as that of the Si epitaxial growth without nitridation (about 3.3Å/min). In our previous work,[5] it was observed with FTIR analysis that the thermal nitridation of Si(100) by NH_3 proceeds with desorption of the hydrogen atom bonded to the Si atom, and the dangling bonds on the Si surface might be covered with the N atoms instead of the hydrogen atoms.[8] From these results, it is suggested that the incubation period is caused by suppression of SiH_4 adsorption and/or reaction on the nitrided site rather than that on the pure Si site.

The Depth Profile of N Atoms in the N Delta-Doped Si Films

The depth profile of the N concentration in the Si film with two N delta-doped regions are shown in Fig. 6. Two N peaks are clearly observed inside Si away from the surface. Each N delta-doped layer can be resolved with the thickness of only 10 Å. The total amount of N atom in each N delta-doped region is approximately $1.4\times10^{14}cm^{-2}$ and almost in agreement with that of the initial atomic-order nitridation. This indicates that the N atoms can be buried at the initial position without N desorption during the capping Si growth.

Thermal Stability and Chemical State of the N Delta-doped Si Films

The depth profile of the N concentration in the $8\times10^{13}cm^{-2}$ N delta-doped Si films with

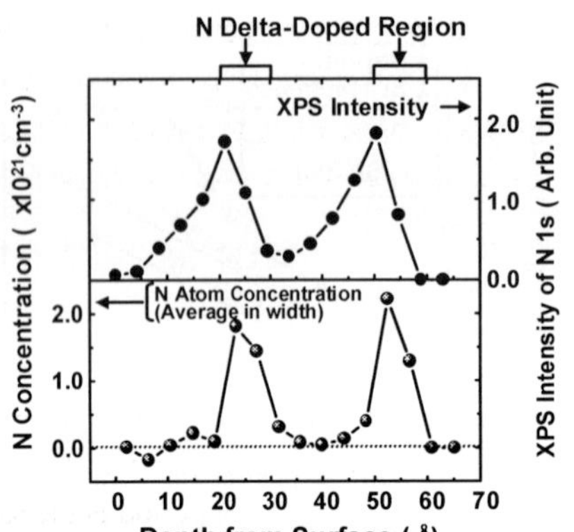

Fig.6 (a) Measured depth profile of the N concentration for the deposited Si film with two N delta-doped regions. The total amount of N atoms is approximately $1.4\times10^{14}cm^{-2}$ for each N delta-doped region. Calculated N concentration will include a detection error around $1\times10^{20}cm^{-3}$.

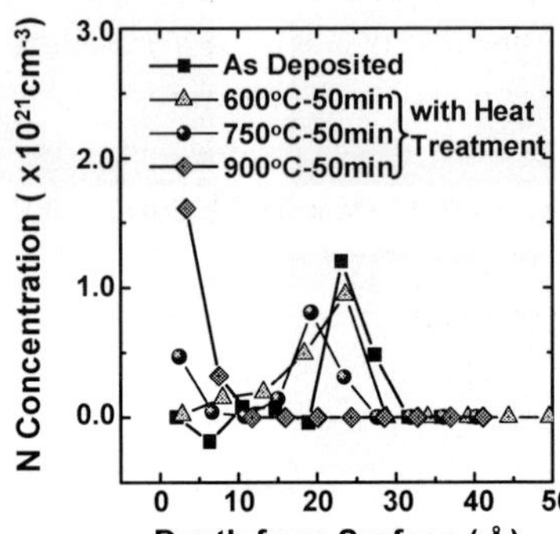

Fig.7 Depth profile of the N concentration for the as-deposited N delta-doped Si film and the samples with the heat treatment at 600, 750 and 900°C. The total amount of N atoms is approximately $8\times10^{13}cm^{-2}$.

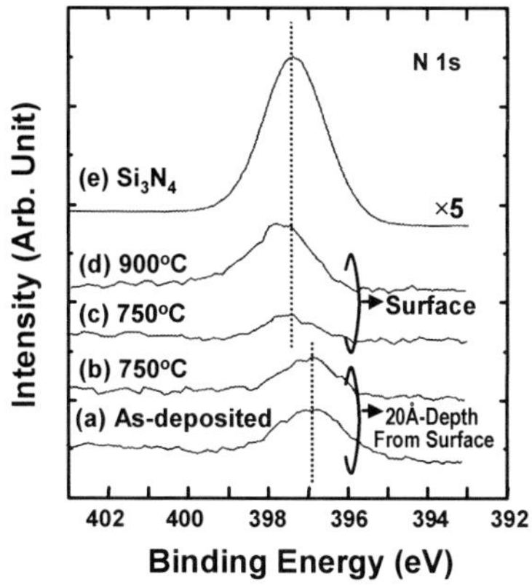

Fig.8 XPS spectra of the N 1s core level measured from (a) the as-deposited N delta-doped Si film and the sample with heat treatment in N_2 ambient at ((b) and (c)) 750°C and (d) 900°C and (e) the Si_3N_4 film deposited with SiH_4 and NH_3 at 800°C on Si(100). The sample surface was etched by 20Å for (a) and (b).

heat treatment is shown in Fig. 7. The concentration in the initial N delta-doped region decreases, while the concentration at the outermost surface increases with the temperature during heat treatment. These results indicate that the N atoms in the delta-doped region tend to diffuse during heat treatment, and especially for the case of thermal treatment at 900°C, most of the N atoms pile up at the surface.

The typical XPS spectra of N 1s from the N delta-doped Si film before and after heat treatment at 750°C and the Si_3N_4 film are shown in Fig. 8. The binding energy of N 1s for the surface agrees with that for Si_3N_4, but a 0.6 eV lower binding energy is observed at the surfaces of the as-deposited and the thermally treated samples etched by 20 Å. It is suggested that N atoms delta-doped in the film have not Si_3N_4 structure.

4.Conclusions

The epitaxial growth of the N delta-doped Si film with the total amount of N atoms as high as $3.0\times10^{14}cm^{-2}$ for a delta-doped region, has been realized at 500°C by ultraclean low-temperature low-pressure CVD. On the atomic-order nitrided Si(100), an incubation period for Si deposition is observed, while the deposition rate after the incubation period is the same as that on the Si surface without the nitridation. After capping Si layer growth, even in the Si film with two N delta-doped regions, most of the N atoms are buried in the initially nitrided region with the thickness of about 10Å. By thermal treatment of the N delta-doped Si film, N atoms in doped region tend to diffuse, and especially for the case of thermal treatment at 900°C, most of the N atoms pile up at the surface. The binding energy of N 1s for the surface agrees with that for Si_3N_4, but a 0.6 eV lower binding energy is observed at the surfaces of the as-deposited and the thermally treated samples etched by 20 Å. It is suggested that N atoms delta-doped in the film have not Si_3N_4 structure.

Acknowledgments

This study was carried out in the Superclean Room of Laboratory for Electronic Intelligent Systems, Research Institute of Electrical Communication, Tohoku University. The CVD reactor system was provided by Hitachi Kokusai Electric Inc.. This study was partially supported by the Public Participation Program for the Promotion of Information Communications Technology R&D from the Telecommunications Advancement Organization (TAO) of Japan, and a Grant-in-Aid for Scientific Research from the Ministry of Education, Culture, Sports, Science and Technology of Japan, and Research for the Future Program in the Area of Atomic-Scale Surface and Interface Dynamics (#JSPS-RFTF97P00202) from the Japan Society for the Promotion of Science.

References

[1] T. Watanabe, M. Sakuraba, T. Matsuura and J. Murota, *Jpn. J. Appl. Phys.* **38**, (1999) 515.

[2] T. Watanabe, M. Sakuraba, T. Matsuura and J. Murota, *J. Vac. Sci. Technol.* A **19**, (2001) 1907.

[3] J. Murota and S. Ono, *Jpn. J. Appl. Phys.* **33**, (1994) 2290.

[4] T. Watanabe, M. Sakuraba, T. Matsuura and J. Murota, *J. Electrochem. Soc.* **145**, (1998) 4252.

Heavy Doping Characteristics of Si Films Epitaxially Grown at 450°C by Alternately Supplied PH_3 and SiH_4

Yosuke Shimamune, Masao Sakuraba, Takashi Matsuura and Junichi Murota

Laboratory for Electronic Intelligent Systems,
Research Institute of Electrical Communication, Tohoku Uniersity
2-1-1 Katahira, Aoba-ku, Sendai 980-8577, Japan
Phone:+81-22-217-5548 Fax:+81-22-217-5548 E-mail:murota@riec.tohoku.ac.

1.Introduction

Atomic-layer doping is one of the important processes in order to fabricate the ultrasmall structure devices. There were some reports of atomic-layer doping by molecular beam epitaxy (MBE) [1]. However chemical vapor deposition (CVD) offers many advantages over MBE, such as high throughput, in-situ doping and selective deposition. Atomic-layer doping of P and B in SiGe epitaxy was achieved using rapid thermal CVD [2]. We have achieved the heavily P-doped epitaxial Si film by alternately supplied PH_3 and SiH_4 at 450°C as shown in Fig.1. In the present work, the P doping process and the thermal stability of the electrical characteristics of the multi-layer P-doped epitaxial Si films have been investigated.

2.Experimental

Epitaxial growth of multi-layer P-doped Si was carried out by alternately supplied PH_3 and

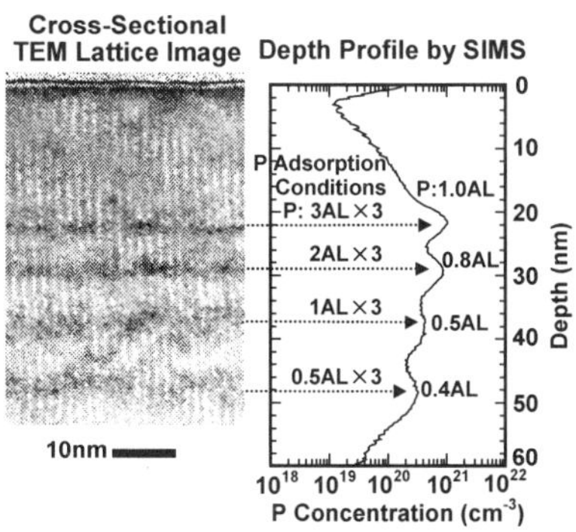

Fig.1 Cross-sectional TEM lattice image and the depth profile of the P concentration of the multi-layer P-doped epitaxial Si film. The fillm is grown by alternately supplied PH_3 at 300-450°C and SiH_4 at 450°C. The 0.5, 1, 2 and 3 atomic-layer formation were done by PH_3 exposure at a partial pressure of 0.26Pa at 300°C for 30min, and at 450°C for 3, 30 and 90min, respectively [3]. SiH_4 pressure was 220Pa. It should be noted that the P concentration shown in ref.4 and 5 was over estimated by using wrong sensitive factor in SIMS measurement.

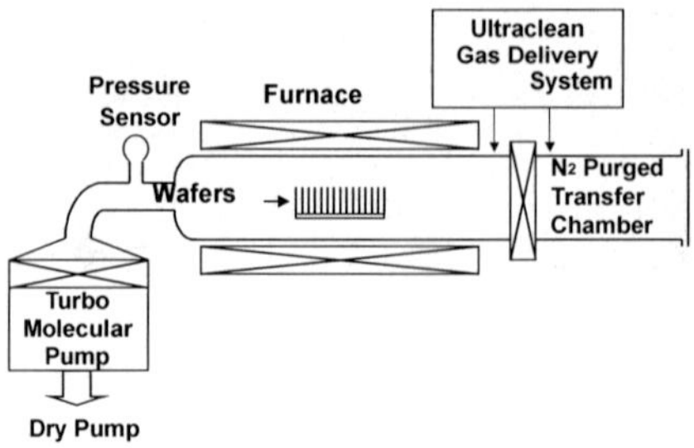

Fig.2 Ultraclean hot-wall low-pressure chemical vapor deposition system.

SiH_4 using an ultraclean hot-wall low-pressure CVD system, schematically shown in Fig.2. This system was made ultrahigh vacuum compatible, the details of which were described elsewhere [6,7].

The substrates used were the mirror polished p-type Si(100) wafers of 8.5–11.5 Ω cm. The substrates were cleaned for several cycles in a 4:1 solution of H_2SO_4 and H_2O_2, rinsed with high purity DI water, dipped in a diluted HF solution for oxide removal, and rinsed with DI water. Then the substrates were loaded into the reactor. A Si epitaxial layer of about 200nm was grown on the Si substrates at 750°C using SiH_4 and H_2 gases to obtain a clean Si(100) surface before PH_3 exposure. After the epitaxial growth of Si, those samples were cooled down to PH_3 exposure temperature of 450°C in Ar. P layers at amount of 1.4×10^{15} and $2.0 \times 10^{15} cm^{-2}$ were deposited on the Si surface by PH_3 exposure at a partial pressure of 0.26Pa for 30 and 90min, respectively [3]. In order to form the multi-layer P-doped Si films(capping Si / P / Si spacer / P ··· P / epi. Si / Si(100) substrate), alternative exposure to PH_3 at 450°C and to SiH_4 at a pressure of 220Pa at 450°C were done.

To determine the surface P amount and the P amount inside Si, angle-resolved x-ray photoelectron spectroscopy (XPS) and secondary ion mass spectroscopy (SIMS) were used, respectively. Crystallinity of the deposited films were observed by reflection high-energy electron diffraction (RHEED). Electrical characteristics of the films were evaluated by the van der Pauw method.

3.Results and discussions

In the 2-layer P-doped epitaxial Si films, with the Si spacer thickness in the range of 2-26nm, no degradation of the crystallinity is observed in the RHEED patterns of those films as shown in Fig.3, although the total P amount of 2 layers decreases with decreasing the Si spacer thickness.

The comparison between P_{2p} XPS intensities detected at the take-off angle (TOA) of 35° and 90°, for the sample exposed to SiH_4 after P layer formation, is shown in Fig.4. The P_{2p} XPS intensities detected at TOA of 35° and 90° decrease with increasing the SiH_4 exposure time. In the case of 22Pa SiH_4 exposure, ratio of P_{2p} XPS intensity at TOA of 35° to that of

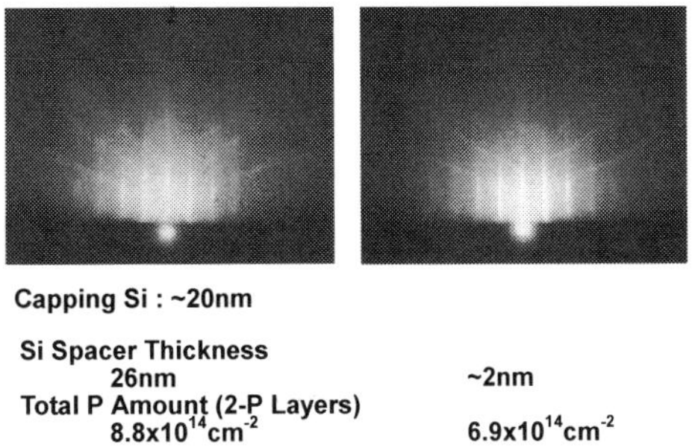

Fig.3 RHEED patterns of the 2-layer P-doped epitaxial Si films with the Si spacer thickness of 26nm and about 2nm. Each film was grown by 2-cycles of alternative exposure to PH_3 and to SiH_4 at 450°C.

90° is constant in spite of the SiH_4 exposure time. This means that P desorbs without Si deposition by SiH_4 exposure. Fig. 5 shows clearly that the surface P amount decreases with increasing the SiH_4 exposure time. Since the thermal desorption of P was scarcely observed at 450°C in H_2 ambient of 1330Pa[3], it is considered that P layer formed on Si surface is reduced by SiH_4. In the case of 220Pa SiH_4 exposure, the ratio of intensity at TOA of 35° to that of 90° increases with increasing the SiH_4 exposure time. This means that Si is grown on the P layer formed surface, although the P may be reduced by SiH_4 or segregates to the capping Si surface. It was reported that Si_2H_6 is produced due to the polymerization of SiH_4 and increases in proportion to the square of the SiH_4 partial pressure [8]. In the case of 220Pa SiH_4 exposure, Si deposition may be enhanced by the Si_2H_6 produced from SiH_4.

The depth profiles of the P concentration in the 2-layer P-doped epitaxial Si films, formed by alternately supplied PH_3 and 220Pa SiH_4, are shown in Fig.6. In the sample (a), with a Si spacer thickness of 26nm, P amount of Layer 1 and Layer 2 are in good agreement with each

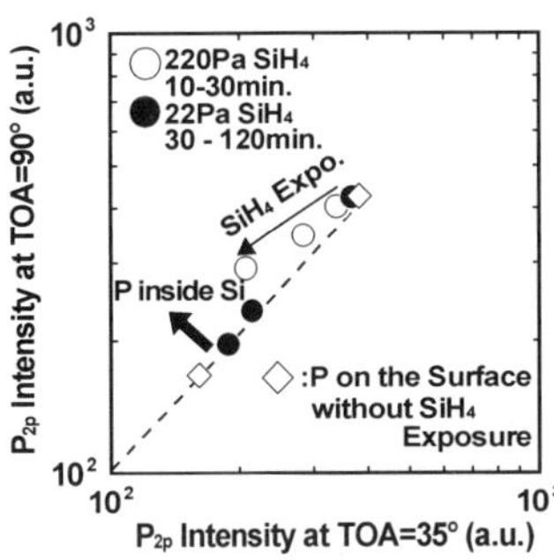

Fig.4 Comparison between P_{2p} XPS intensity detected at the take-off angle of 35° and 90°. Plots on the dotted line means that P atoms exist on the surface. Surface P amount before SiH_4 exposure is $2x10^{15}cm^{-2}$.

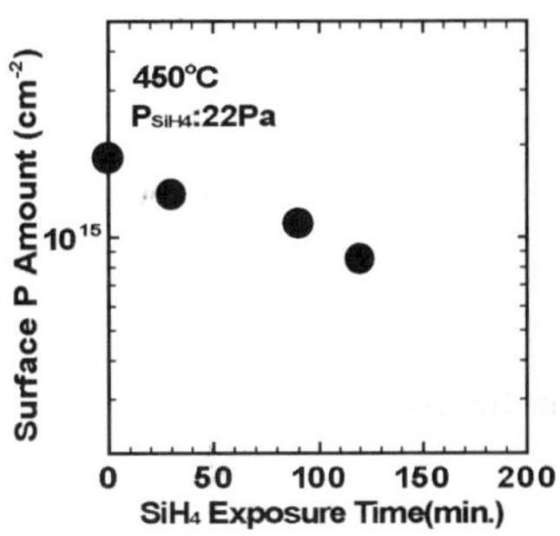

Fig.5 SiH_4 exposure time dependence of the surface P amount on Si. These data are derived from the P_{2p} XPS intensity as shown in Fig.4. Surface P amount before SiH_4 exposure is $2x10^{15}cm^{-2}$.

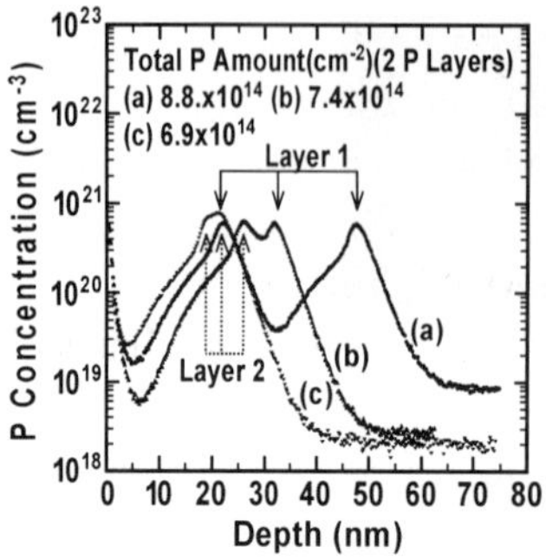

Fig.6 The typical depth profiles of P concentration in the 2-layer P-doped epitaxial Si films.

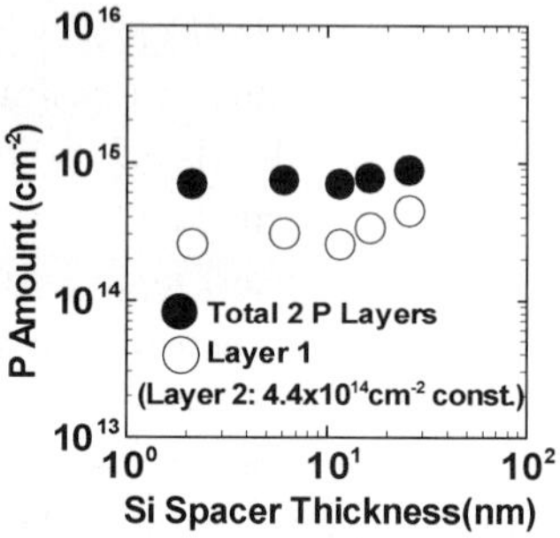

Fig.7 Si spacer thickness dependence of P amount in the 2-layer P-doped epitaxial Si films. Amount of Layer 1(first P doping) is calculated with assumption of that amount of Layer 2 (second P doping) is constant of $4.4x10^{14}cm^{-2}$ in spite of Si spacer thickness.

other ($4.4x10^{14}cm^{-2}$), although the P atoms distribute due to the resolution of SIMS measurement, as shown in the P tail towards to substrate side. Because surface P amount before Si deposition is $2.0x10^{15}cm^{-2}$, it is considered that the P atoms of about $1.6x10^{15}cm^{-2}$ desorb from the Si surface during Si deposition and P atoms of $4.4x10^{14}cm^{-2}$ are incorporated into capping Si or Si spacer from the existence of P tail towards to surface side.

It is found that the total P amount decreases slightly with decreasing the Si spacer thickness as shown in Fig.7. In the case of thinner Si spacer thickness below 20nm, it can be assumed that the P segregates to the surface of Si spacer from the P formed layer, and in the second P doping (Layer 2 in Fig.6.), P adsorbs on the P segregated surface of Si spacer. Therefore, the P atoms incorporated into capping Si must be $4.4x10^{14}cm^{-2}$, because capping Si thickness is thicker than 20nm. Then, the P amount in Layer 1 decreases with decreasing the Si spacer thickness. By extrapolating the amount of Layer 1 at the Si spacer thickness of Si

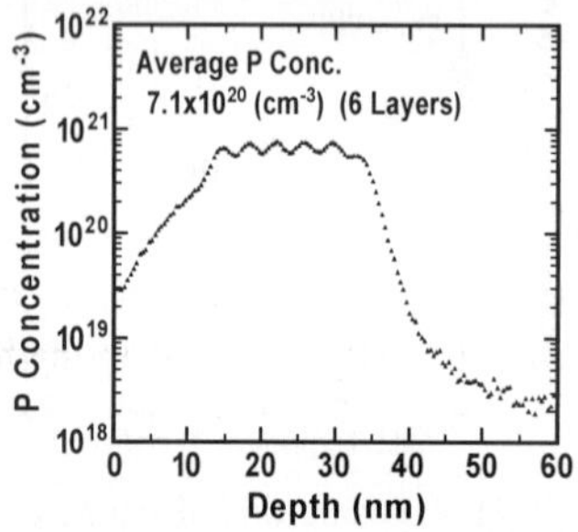

Fig.8 Depth profile of the P concentration in the multi-layer P-doped epitaxial Si film. Each P layer formation condition is surface P amount of $1.4x10^{15}cm^{-2}$. Si spacer thickness is 3.7 nm.

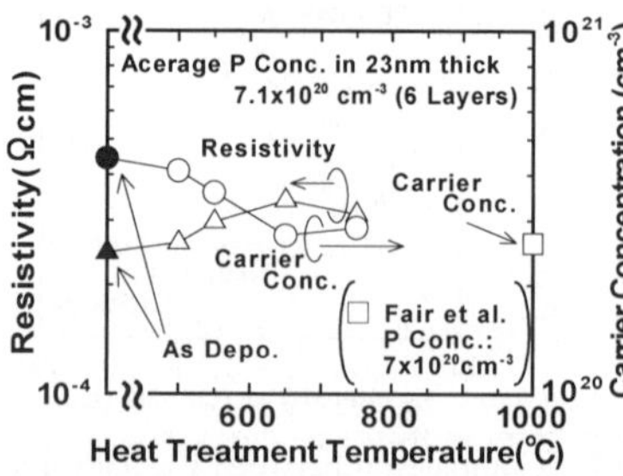

Fig.9 Heat treatment temperature dependence of the resistivity and the carrier concentration of the multi-layer P-doped epitaxial Si film, as shown in Fig.8. Treatment time is 60min.

atomic-layer (0.14nm) in the Fig.7, it is considered that the amount of P, existing on the initial surface without incorporation into Si spacer, must be $1\sim2\times10^{14}cm^{-2}$ order.

Typical depth profile of the multi-layer P-doped epitaxial Si film, with average P concentration of $7\times10^{20}cm^{-3}$ and Si spacer thickness of 3.7nm, is shown in Fig.8. It is found clearly that 6 layers of P are doped in the film. In such heavily P-doped Si film, 60% of P is electrically active and the resistivity is $2.4\times10^{-4}\Omega cm$. By heat treatment of the film at above 550°C, the carrier concentration decreases and the resistivity increases, as shown in Fig.9. It was reported that electrically inactive SiP precipitates are formed in Si doped with P ($7.5\times10^{19}cm^{-3}$) by heat treatment at above 500°C for 500hours [9]. In the present case of the multi-layer P-doped epitaxial Si film, with high P concentration beyond $10^{20}cm^{-3}$, it is considered that SiP precipitates may be formed at above 550°C even for a short period of 60min. After the heat treatment at above 550°C, the carrier concentration and the resistivity of the film become close to those of the P-doped epitaxial Si film, with P concentration of $7\times10^{20}cm^{-3}$, formed by P diffusion from $POCl_3$ source at 1000°C [10]. It is considered that more electrically active P atoms exist in the heavily P-doped Si film formed at 450°C compared with those in the P-doped Si films formed at higher temperature at the same P concentration. In the case of heat treatment of the Si epitaxial film, consisting of the heavily P-doped layer with P concentration of $\sim10^{21}cm^{-3}$ with a Si spacer thickness below 1nm, at above 550°C, resistivity increased up to 4 times higher than that before the heat treatment and the degradation of crystallinity was observed [4]. It is considered that, in the heavier P-doped Si film, the more SiP precipitates are formed by heat treatment. These results suggest that the very low-resistive heavily P-doped epitaxial Si films can be achieved at very low-temperature around 450°C.

4.Conclusions

On the P formed layer surface, Si deposition is achieved by 220Pa SiH_4 at 450°C although the P reduction by SiH_4 occurs. In the 2-layer P-doped epitaxial Si films, P atoms incorporated into Si spacer decrease with decreasing the Si spacer thickness. It is suggested that P segregates to the surface of Si spacer from P layer formed surface during Si deposition. As a result, P of $1\sim2\times10^{14}cm^{-2}$ is considered to be on the initial surface. By alternately supplied PH_3 and SiH_4 at 450°C, very low-resistive heavily P-doped epitaxial Si film is achieved. By heat treatment of the film at above 550°C, the carrier concentration decreases and the resistivity increases. It is considered that more electrically active P atoms exist in the heavily P-doped Si film formed at 450°C compared with those in the P-doped Si films formed at higher temperature at the same P concentration. These results suggest that the very low-resistive heavily P-doped epitaxial Si films can be achieved at very low-temperature around 450°C.

Acknowledgments

This study was partially supported by the Public Participation Program for the Promotion of Info.

Communications' Technology R&D from the Telecommunications Advancement Organization of Japan, and the Research for the Future Program (no. JSPS-RFTF97P00202, Atomic-Scale Surface and Interface Dynamics) from Japan Society for the Promotion of Science and a Grant-in-Aid for Scientific Research from the Ministry of Education, Culture, Sports, Science and Technology of Japan. The CVD reactor system was provided by Kokusai Electric Co., Ltd.

References

[1] K.Nakagawa, A.A.van Gorkum and Y.Shiraki, *Appl. Phys. Lett.*, **54**, (1989) 1869.
[2] B.Tillack, *Thin Solid Films,* **318**, (1998) 1.
[3] Y.Shimamune, M.Sakuraba, T.Matsuura and J.Murota, *Appl. Surf. Sci.*, **162-163**, (2000) 390.
[4] Y.Shimamune, M.Sakuraba, T.Matsuura and J.Murota, *Thin Solid Films*, **380**, (2001) 134.
[5] Y.Shimamune, M.Sakuraba, T.Matsuura and J.Murota, *J. de. Physique IV*, **11**, Pr3 (2001) 255.
[6] J.Murota, N.Nakamura, M.Kato, N.Mikoshiba and T.Ohmi, *Appl. Phys. Lett.*, **54**, (1989) 1007.
[7] J.Murota and S.Ono, *Jpn. J. Appl. Phys.*, **33**, (1994) 2290.
[8] T.Morie and J.Murota, *Jpn. J. Appl. Phys.*, **23**, (1984) L482.
[9] P.Ostoja, D.Nobili, A.Armigliato, R.Angelucci, *J.Electrocehm.Soc.*, **123**, (1976) 124.
[10] R.B.Fair and J.C.C.Tsai, *J.Electroceh. Soc.*, **124**, (1977) 1107.